SpringerBriefs in Animal Sciences

For further volumes:
http://www.springer.com/series/10153

Bina Pani Das

Mosquito Vectors of Japanese Encephalitis Virus from Northern India

Role of BPD Hop Cage Method

 Springer

Bina Pani Das
Department of Biosciences
Jamia Millia Islamia
New Delhi
Delhi
India

ISSN 2211-7504   ISSN 2211-7512   (electronic)
ISBN 978-81-322-0860-0   ISBN 978-81-322-0861-7   (eBook)
DOI 10.1007/978-81-322-0861-7
Springer New Delhi Heidelberg New York Dordrecht London

Library of Congress Control Number: 2012949707

Printed on acid-free paper

Springer is part of Springer Science+Business Media (www.springer.com)

Dedicated to
My husband (Lalit Mohan Das)
Daughters (Madhumita and Sangeeta)
Sons-in-Law (Sunil and Avinash)
Son (Dibyendu)
Grand children (Kalyani and Aniket)

Foreword

JAMIA MILLIA ISLAMIA
(A Central University by an Act of Parliament)

Najeeb Jung, *IAS*

Vice-Chancellor

Maulana Mohamed Ali Jauhar Marg, New Delhi-110025   Tel. : 26984650, 26985180   Fax. 00-91-11-26981232
Email: vc@jmi.ac.in   website : http://www.jmi.nic.in

Foreword

Japanese encephalitis (JE) is transmitted by mosquitoes and has now emerged as a major viral disease frequently occurring in epidemic manifestation in many parts of India.

The author of this book, Dr. Bina Pani Das, has been associated with mosquito vectors of human diseases since 1980s; first with Indian Council of Medical Research as Research Associate and then for over a period of 22 years with National Institute of Communicable Diseases (NICD), DGHS, Government of India, from where she retired in 2007 as Joint Director.   Currently she is with the Jamia Millia Islamia in connection with "Mosquito vectors of Japanese encephalitis virus from Northern India" a project supported by Department of Science and Technology, Govt. of India.

She has published books on Vespid Fauna of the Indian Subregion; Catalogue of species belonging to families of Social Wasps covering the Indian Subregion; and rapid identification of Anopheline mosquito species so far known from India. Her last book, "Pictorial Key to the Species of Indian Anopheline Mosquitoes" published in 1990 has been distributed by the World Health Organization, SEARO, New Delhi in some South East Asian Countries where the mosquito fauna is more or less similar. The book has been widely acclaimed and proved useful reference for rapid identification of Anopheline mosquitoes in the region.

During the last ten years (1998-2007) at NICD, Dr. Das had focused on entomological aspect of Japanese encephalitis. Within the initial phase (1999), she discovered a protozoan bio-control agent killing JE vector larvae growing in paddy fields. This has been done first time in science and Dr Das as the Inventor was granted International patents.

M. A. Zaki
1997-2000

Syed Shahid Mahdi
2000-04

Mushirul Hasan
2004-09

The present book is aimed at establishing entomological evidence for Japanese encephalitis outbreak in Northern India during nine years of field oriented research on ecology of JE vectors from North India. It is on reading the book that one realizes the tireless effort that has gone unto its making. Collating and analysing enormous field as well as laboratory based relevant research on JE vectors undertaken in various parts of our country must have been no mean task. I believe there would hardly be another book of this kind and it will prove to be a useful handbook for medical entomologists and students, especially those involved in mosquito research and control. I congratulate Springer for their foresight in commissioning this book and wish Dr. Das all success in her future endeavours.

With pride, I state that Jamia is proud to have Dr.Das with us.

Najeeb Jung (IAS)
Vice-Chancellor

Preface

Japanese Encephalitis (JE), transmitted mostly by *Culex tritaeniorhynchus* belonging to *Culex* vishnui group of mosquitoes, is the leading cause of viral encephalitis in 14 Asian countries. Approximately, 60 % of the world's population live at risk in JE endemic regions of these countries. In humans, JE virus causes inflammation of the membranes around the brain hence the name "encephalitis".

JE started affecting India since 1970s and now it has emerged as a major public health problem due to its epidemic potential, high case fatalities and lifelong disabilities in survivors. Many major JE outbreaks have been reported from different parts of the country and in most of these outbreaks from Northern India, disease transmission could not be explained due to negligible vector density detected during entomological investigations. Every year, so-called "undiagnosed viral illness" invades India and unfailingly claims thousands of lives especially in children below 15 years. Of these several hundred child deaths, >75 % are contributed by Northern India with case fatality rate ranging from 10 to 77.5 %.

Culex. tritaeniorhynchus, primary vector of JE in India is predominantly exophilic i.e. rest outdoors and normally zoophilic, i.e. they prefer to take blood meals from animals. JE virus circulates and multiplies among pigs and birds and infects these zoophilic mosquitoes. These mosquitoes turn into indiscriminate feeder, increasing man–vector contact leading to transmission of JE virus in man during monsoon and post-monsoon months when their density increases tremendously. Therefore, the mosquito sampling techniques used need to be adequately sensitive to detect the sharp increase in vector density for initiating integrated vector control measures to prevent JE outbreak. However, the main reason for failure in detecting sharp increase in JE vector density in earlier JE outbreak investigations from Northern India was the use of inadequate mosquito sampling tools.

In order to overcome the above problem, I have developed "BPD Hop Cage Method" a simple, cost-effective, operationally feasible sampling tool specially designed to capture predominantly outdoor resting mosquitoes from land and aquatic vegetation. This has helped to study nearly every aspect of JE vector bionomics and establish entomological evidences of JE outbreaks occurring in Northern India upon its use since 2003.

This book mainly includes data generated by me on ecological studies of JE vectors undertaken from Northern India over a period of 9 years (1998–2006) at

National Institute of Communicable Diseases (NICD), Delhi, and supported by data collated during outbreak investigations of JE/Acute Encephalitis Syndrome (AES) carried out in different parts of the country by central teams (constituted by the Ministry of Health and family Welfare) in which I was associated as a member. Based on the observation obtained, situation specific integrated vector control strategies were suggested to prevent transmission of the disease in Northern India. Similar JE vector surveillance and vector control measures are suggested for other regions of South East Asia where similar ecological and environmental conditions exist.

Health including control of mosquito borne diseases is a state subject. Entomological man power and set-up available with the state authorities are either very poor or does not exist. In order to improve availability of trained man power in the country, students of Master of Public Health, IP University, Delhi and in-service public health personnel of India and abroad undergoing various training programmes held at NICD were taken for field studies to the JE endemic areas in Karnal district (Haryana) and Saharanpur district (Uttar Pradesh). They were exposed to the operational methodology of various sampling tools used in detecting JE vector abundance during outbreak investigation of JE/AES, along with other procedures for collecting data on environmental and epidemiological parameters of the disease.

Though JE virus cannot be eliminated from the environment, as it is not possible to kill all the infected reservoir birds; however, the disease burden can certainly be reduced appreciably by efficient assessment of JE vector abundance and JE virus infection in local vector mosquitoes. A prime requisite for this is the accurate determination of the species of *Culex* mosquitoes actually involved in transmitting the disease. The only available key can be used by those familiar with taxonomic language and not by common users in the programme. I present here a simple illustrated key, in a language which is tuned to the medical officers and paramedical staff in public health programme, to differentiate 17 commonly encountered species of *Culex* (*Culex*) mosquitoes associated with JE in India.

The objective of the book is to disseminate the knowledge gained by me over a period of nearly last 15 years of research in the field of ecology of mosquito vectors of JE virus from Northern India to anyone who wishes to curtail death of children due to this dreaded disease. I urge you to send me your suggestions for improving the book further.

There are many people behind the successful completion of this book, without whose help I could not have brought this assignment to fruition. I thank Dr. N. Arunachalam, Centre for Research in Medical Entomology (CRME), Madurai, India and Prof. V. K. Gupta, University of Florida, USA for their critical comments on some chapters of the manuscript; Mr. N. L. Katra, former Entomologist, National Institute of Communicable Diseases (NCID) for his valuable suggestions which enriched this book. My sincere gratitude is due to the successive Directors and especially to Dr. V. K. Saxena, Head, Centre for Medical Entomology & Vector Control, National Centre for Disease Control (formerly NICD) for facilities during the course of experiments earlier at NICD. I would like to thank my staff members at NICD for their assistance in the field and laboratory studies. I express my sincere thanks to successive Officers-in-Charge, CRME, for allowing me to learn and utilise the facilities available at their institute for JE virus antigen detection; Dr. A. P. Dash, former Director and Dr. Sukla Biswas, National Institute for Malaria Research for providing technical assistance with the mosquito blood meal identification. I am also thankful to Mr. H. C. Agrawal, Director, Postal Training College, Saharanpur for graciously providing space to establish my field laboratory every month from July 2005 to June 2006 in his institute; Dr. O. P. Singh, D. M. O, Saharanpur and Mrs. Bamba, Biologist, Karnal, for all the facilities and help extended during field work at the respective districts.

I am grateful to the Department of Science and Technology (DST) as the project is catalysed and supported under its Utilisation of Scientific Expertise of Retired Scientist Scheme. I would like to thank Dr. Rita Banerjee, DST, for encouragement. I must express my most profound gratitude to my mentor Prof. Syed Akhtar Husain, Jamia Millia Islamia (JMI) for his expert knowledge, invaluable suggestions and unstinted support during the course of compilation of this book. Sincere thanks are also due to Prof. L. Khan, Head and Dr. Amit Kumar In-charge, Central Instrumentation Facility (CIF), Department of Biosciences (JMI) for providing the laboratory facilities during the course of this project. I must acknowledge the hard work and dedication shown by Mrs. Karuna Patil, CIF, Arif Tasleem Jan, Mudsser Azam and Md. Salman Akhtar, Ph.D. students for their assistance in maintaining and improving my culture strain at JMI.

I am blessed with a wonderful family who, over the years, has proved to be my most reliable team members. I thank all of them from the bottom of my heart.

Bina Pani Das

Former Joint Director, NICD, Delhi
Jamia Millia Islamia, New Delhi, India

About the Book

Japanese Encephalitis (JE), a mosquito borne disease, is the leading cause of viral encephalitis in 14 Asian countries due to its epidemic potential, high case fatality rate and increased possibility of lifelong disability in patients who recover from this dreadful disease. In spite of seriousness of the disease, still only few books are available for ready reference. Hence, this book will be useful for students, entomologists, paramedical staff and vector control managers in public health.

The objective of the book is to disseminate the knowledge gained by the author from ecological studies on JE vectors undertaken in two endemic and two non-endemic areas of Northern India over a period of last 15 years (1998–2012) of research in the field of ecology of mosquito vectors of JE virus initially at National Institute of Communicable Diseases (Ministry of Health & Family Welfare, Government of India), Delhi and later at Jamia Millia Islamia, a Central University, Delhi, to anyone who wishes to curtail death of children due to this dreaded disease.

Of the thousand suspected JE deaths in India annually, more than 75 % are contributed by Northern India wherein disease transmission failed to be explained based on entomological evidence due to inadequate mosquito surveillance tool used in determining JE vector density. In order to overcome the above problem, Dr. Bina Pani Das, the author of this book, developed "BPD Hop Cage Method", a simple, cost effective and operationally feasible surveillance tool specially designed to capture predominantly day resting adult *Cx. tritaeniorhynchus* mosquitoes, the principal JE vector species in the country from land and aquatic vegetation.

About the Author

 Dr. Bina Pani Das Ph.D., is a Medical Entomologist and Former Joint Director, NICD, Delhi wherein she was associated from 1985 to 2007 in field oriented research related to mosquito borne diseases of great public health concern like malaria, dengue/Chikungunya and Japanese encephalitis. At present, she is with Jamia Millia Islamia University, Delhi, in connection with "Mosquito vectors of Japanese encephalitis virus from Northern India" a project supported by Department of Science and Technology, Government of India.

The author has over 30 years of proven expertise in preparation of catalogues, check lists and pictorial keys of insects belonging to social wasps (Hymenoptera: Vespidae and Stenogastridae) and mosquitoes (Diptera: Culicidae).

During field studies, the author has discovered pathogenic property of a ciliate (microbe) first time in science. Her passion for research led her to become the inventor of the patent entitled "Microbial control agent for mosquito vector of human diseases" which has so far been granted by six countries, viz. USA, Australia, Sri Lanka, Vietnam, Bangladesh and Philippines.

The author has developed "BPD Hop Cage Method"—a simple, cost effective and programme oriented technique for effective JE vector surveillance.

Contents

Abbreviations

AES	Acute Encephalitis Syndrome
ATCC	American Type Culture Collection
BPD	Bina Pani Das
CFR	Case fatality rate
CIF	Central instrumentation facility
CRME	Centre for Research in Medical Entomology
CSF	Cerebro spinal fluid
DC	Dusk collection
DST	Department of Science and Technology
EIA	Enzyme immuno assay
ELISA	Enzyme linked immuno sorbent assay
HC	Hand collection
IDA	International Depositary Authority
IgG	Immunoglobulin G
IgM	Immunoglobulin M
JE	Japanese Encephalitis
JEV	Japanese Encephalitis Virus
JMI	Jamia Millia Islamia
MRC	Malaria Research Centre
NCDC	National Centre for Disease Control
NICD	National Institute of Communicable Diseases
NIMR	National Institute of Malaria Research
NIV	National Institute of Virology
NVBDCP	National Vector Borne Disease Control Programme
PMH	Per man hour
PMHD	Per man hour density
PRD	Per room density
PTHC	Per ten hop cages
WHO	World Health Organization

Chapter 1
Japanese Encephalitis and Problem in Vector Surveillance: An Introduction

Abstract Japanese Encephalitis (JE), a mosquito borne disease, is the leading cause of viral encephalitis in 14 Asian countries. Approximately, 60 % of the world's population live at risk in JE-endemic regions of these countries. JE is responsible for several hundred deaths in children annually in India, of which >75 % are contributed by Northern India. The disease was first recognised in 1955 and since then many major out-breaks have been reported from different parts of the country, predominantly in rural areas. Case fatality rate has ranged from 10 to 77.5 % and in patients who recover; complication may lead to lifelong disability. In most of these outbreaks from northern India, disease transmission could not be explained in the past due to inadequate mosquito surveillance tool used for determining JE vector density. This chapter briefly traces history of emergence of JE in India and provides a general introduction to "Japanese encephalitis". The health impact in construction of two canals, viz. Eastern Yamuna Canal and Western Yamuna Canal in Northern India has also been addressed. The problem faced during outbreak investigation of JE/acute encephalitis syndrome (AES) in Northern India is discussed. Author of this book developed a simple, cost effective, operationally feasible JE vector surveillance tool that helped to know all most every aspect of *Culex tritaeniorhynchus* mosquitoes, principal vector of JE virus in India, and to suggest appropriate vector control measures. These findings are incorporated in Chaps. 2, 3, 4, 5, 6, 7 of this book.

1.1 Introduction

Japanese encephalitis (JE) is an acute, vector borne, noncontagious, and zoonotic viral (flaviviros) infection of the central nervous system (CNS). Over the last three decades, the disease poses a serious public health problem in India because of its increasing frequency of outbreaks, often transforming into epidemics in many parts of the Indian subcontinent and high case fatality rate (CFR) ranging from 10 to 77.5 %. The low case fatality rate may be due to improved hospital facilities and symptomatic treatment. The disease mainly affects the lower socio-economic rural community and in patients who survive, complication may lead to lifelong

B. P. Das, *Mosquito Vectors of Japanese Encephalitis Virus from Northern India*, SpringerBriefs in Animal Sciences, DOI: 10.1007/978-81-322-0861-7_1, © The Author(s) 2013

Map 1.1 Distribution of
Japanese encephalitis

disability especially in children between 1 and 15 years in endemic areas. Adult
infection most often occurs in areas where the disease is newly introduced. In
India, though the disease burden could not be estimated due to poor JE surveillance system, there was an obvious rise in JE incidence in recent years. For every
symptomatic JE case, there are likely to be about 300–1,000 people infected with
JE virus but without any clinical manifestation (asymptomatic cases). There is no
specific treatment for JE.

1.2 Global Scenario

JE is the leading cause of viral encephalitis in South East Asia. Approximately,
60 % of the world's population live in JE-endemic regions and 30,000–50,000
cases and 10,000 deaths per annum were notified from a wide geographic range.
JE is endemic in India, China, Japan, and all of South East Asia (Map 1.1). JE
virus (JEV) was first isolated in Japan in 1935. Vaccination programmes, increased
living standards, and mechanisation of agriculture are key factors in the decline
in the incidence of this disease in Japan and South Korea. However, transmission
of JE is likely to increase in Bangladesh, Cambodia, Indonesia, Laos, Myanmar,
North Korea, and Pakistan because of population growth, intensified rice farm-
ing, pig rearing, and the lack of surveillance system and sustained vaccination
programme.

1.3 Emergence of JE in India

In India, JEV activity was first detected in 1952 through seroepidemiological surveys in Nagpur district of Maharashtra and Chingleput district of Tamil Nadu. JE was clinically recognised for the first time in India in 1955 when cases of JE occurred in Vellore district, Tamil Nadu and Pondicherry in South India. JEV was isolated from the wild caught Culex vishnui group of mosquitoes from the same area of Vellore District in 1956 followed by three isolations from the brain tissues of human cases in 1958. These evidences served to confirm JE as a cause of encephalitis in India. Prior to 1970, cases of JE were reported only from Southern India (Carey et al. 1969). JEV infections were reported in serological surveys carried out between 1955 and 1972 from scattered areas in Gujarat, Maharashtra, Orissa, Assam, and Arunachal Pradesh and highest prevalence in three states of Andhra Pradesh, Karnataka, and Tamil Nadu. In 1973, West Bengal faced the first major epidemic of JE in the country in Burdwan and Bakura district with over 700 cases and 300 deaths followed by a second epidemic in 1976. These were followed by 1978 large-scale epidemic of JE in Dibrugarh district in Assam, Dhanbad district in Bihar, and Gorakhpur district in eastern Uttar Pradesh (UP). Subsequently, the disease spread to other states and has caused a series of outbreaks in different parts of the country.

In Northern India, eastern UP faced a devastating epidemic in October 1978 affecting nearly 3,000 persons (Mathur et al. 1982). The densely populated Terai area of eastern UP bordering Bihar state and Nepal was affected. In Terai area, floods are an annual feature giving rise to water logging for nearly 3–4 months and creating enormous vector population during post flood period. The climate is warm and humid with the result initially the area was endemic for malaria and filaria (Sarkari et al. 1984), but later on became endemic for JE (Khare 1984; Saxena et al. 1986; Narasimham et al. 1988). The disease is endemic in the north eastern districts of UP, viz. Gorahkpur, Deoria, Ajamgarh, Basti, Gonda, Balia, Faizabad, and Lakhimpur Khiri (Chatterjee and Banerjee 1975; Narasimham et al. 1988). JE epidemics have been reported from eastern UP, Bihar, Assam, Manipur, West Bengal, Andhra Pradesh, Tamil Nadu, Karnataka, Goa, and Pondicherry (Banerjee 1988). These were the major rice growing areas of the country. Subsequently, JE has spread to new areas with extensive paddy cultivation like Karnal district in Haryana (1990) and Saharanpur district in UP (2002) in Northern India. These new areas fall in the hot and dry climatic zone with insufficient monsoon rain. They became receptive to JE due to man-made environmental changes by construction of irrigation system.

India is essentially an agricultural country where irrigation has played a very important role in the rural economy in different parts of the country. There are three main types of irrigation in India, viz. canal irrigation, tank irrigation, and well irrigation. While tank irrigation is prevalent in Southern India, well irrigation is extensively used in Rajasthan in Western India, where climate is hot and dry. Canal irrigation is mostly constructed in Northern India where rivers have flow of water

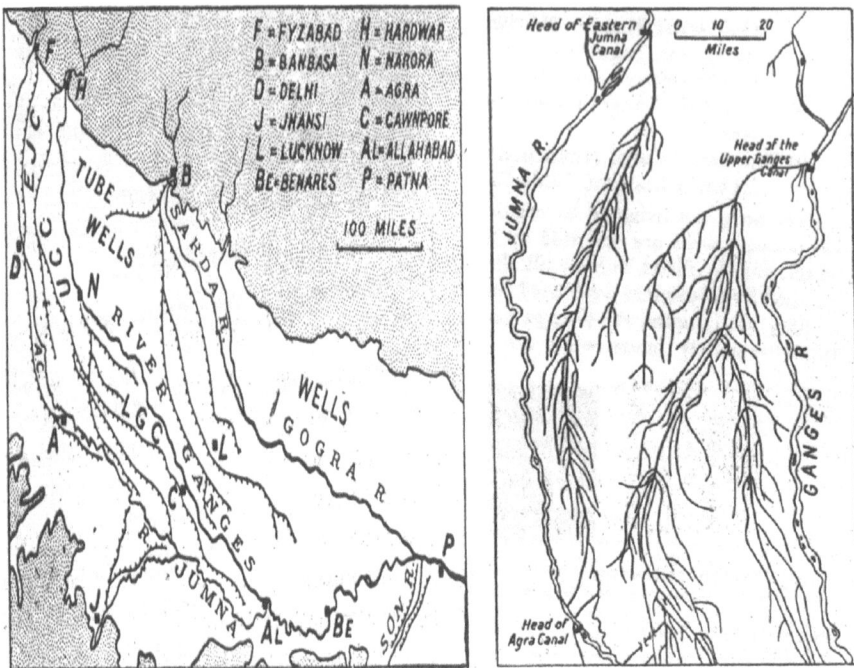

Fig. 1.1   Canal irrigation in Northern India. Eastern Yamuna Canal (EJC)

throughout the year (Das Gupta 1949). Two river systems were harnessed, first was
the River Ganges in 1854 and later the River Yamuna. Two canals were constructed
out of Yamuna, viz. The Eastern Yamuna Canal and Western Yamuna canal. The
Eastern Yamuna Canal takes water from the Yamuna River near Faizabad and irrigates Saharanpur, Muzaffar Nagar, and Bagpat district of north eastern part of UP
(Fig. 1.1). The Western Yamuna Canal irrigates many district of southern Punjab
now called Haryana state. In the absence of any health impact studies, the irrigated
lands in the command area of these two canals became marshy and water logged.
In the initial phases, these command area became highly endemic for malaria and
in the later phase (now) these areas became receptive for JE.

1.4 Magnitude of JE Problem in India

The Directorate of national vector borne disease control programme (NVBDCP)
is monitoring JE incidence in the country since 1978. The data on JE (NVBDCP)
show that a total 101,137 and 33,202 deaths (CFR 32.82 %) due to JE have been
reported in India since 1978 till 2007 (P) from all over the country. The annual
reported incidence for the country has ranged from 1,243 to 7,500. In the year
1978, JE cases were reported from 21 states/union territories.

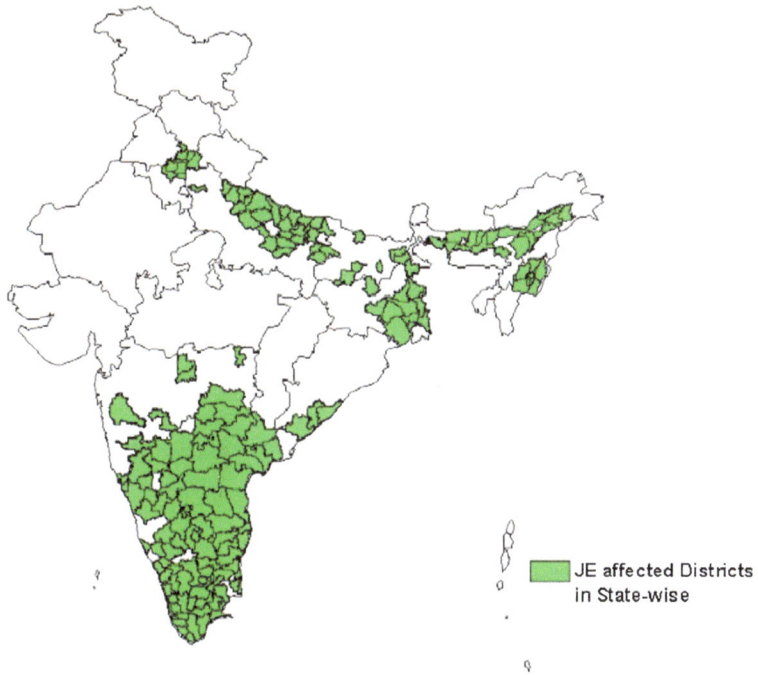

Map 1.2   JE endemic areas of India (2005)

1.4.1 Trend of JE Occurrence

JE has rapidly engulfed a vast area of more than 135 districts in 16 states and Union Territories of the country (Map 1.2). Though cases of JE have been reported from 16 states/UT's in some year(s) since 1978, only ten States (Uttar Pradesh, Andhra Pradesh, Karnataka, Assam, Bihar, Haryana, Tamil Nadu, Goa, West Bengal, and Manipur) are reporting JE regularly. In the past, several epidemics have occurred and the disease is spreading to new areas. During 1955–1965, a total of 52 cases were reported in India. During 1966–1975, reported cases numbered 763 with 325 deaths. From 2001 to 2005 annual cases in the country ranged from 1,695 to 6,587 (Fig. 1.2). In 2005, more than 30 districts of UP were in the grip of JE epidemic since August 2005. Vast areas of more than 135 districts in 12 states of the country are now endemic for JE.

1.4.2 Contribution of JE Cases by States

In 2007, maximum number (75 %) of the AES/JE cases of the country is contributed by UP followed by Assam (10.40 %), Bihar (8.31 %), Karnataka (1.81 %), Haryana and Tamil Nadu 0.75 % each, only 2.98 % cases are contributed by

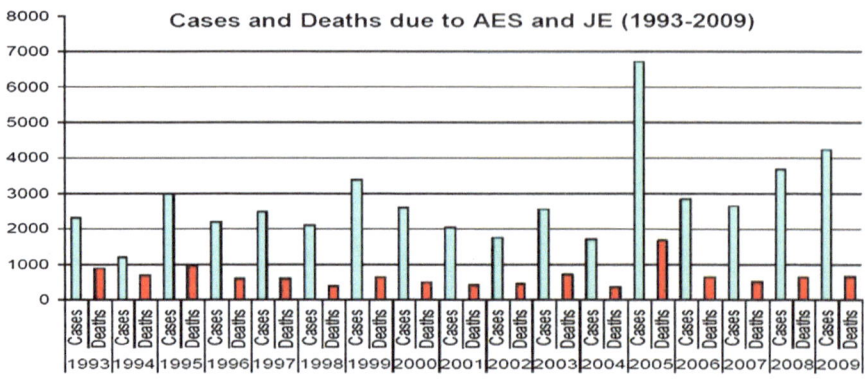

Operational Guide for Japanese Encephalitis Vaccination in India, MoHFW, September 2010

Fig. 1.2 Trend of JE/AES cases and deaths 1993–2009) (*Source* NVBDCP)

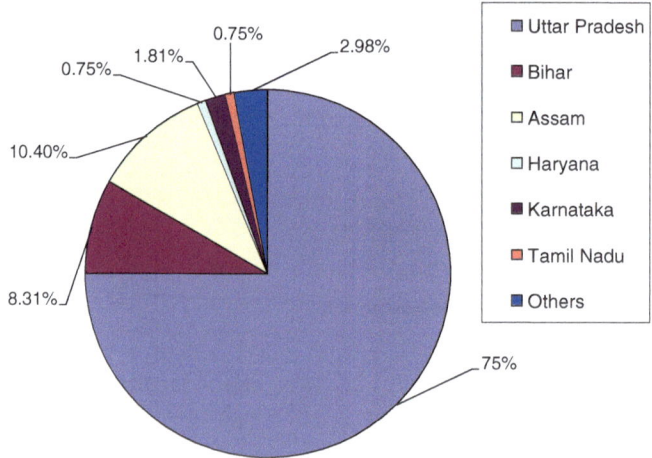

Fig. 1.3 Contribution of AES/JE cases by the states (2007)

remaining six endemic states, viz. Andhra Pradesh, Kerala, Maharashtra, Manipur, West Bengal and Nagaland (Fig. 1.3). Changing pattern of agriculture to rice culti-vation from other crops seems to be one the important determining factor for such rising trend of the disease.

1.4.3 Concept of Endemic and Epidemic Districts in India

In India, for any public health delivery, 'districts' are the administrative unit. Sabesan et al. (2008) for the first time classified "endemic" and "epidemic" form of JE in India on the basis of at least one occurrence of JE during 53 years of observation (1955–2007). According to this classification "Endemic" JE may be defined as "Occurrence of local transmission of JE in an area repeatedly or period-ically at an interval of 3–5 years that remain at a fairly stable prevalence". In con-trast, "Epidemic" JE may be defined as "Unusual occurrence of JE transmission in old foci or fresh occurrence in a newer area and generally with low incidence". In JE endemic areas of south India sporadic cases of JE occur round the year due to congenial climatic condition throughout the year. Of the existing 593 dis-tricts in India 175 (29.5 %) reported JE occurrence at least once during the period 1955–2007. Out of 175 districts, 80 fall in "endemic" category and the remaining 95 districts were classified as "epidemic". Moreover, 85 % of endemic districts experienced up to 11 occurrences, whereas 87 % of epidemic districts experienced up to six occurrences only. Among 80 endemic districts, UP state alone contrib-uted 54 (68 %) of total endemic districts in India (Sabesan et al. 2008). According to this system of classification, Karnal district (Haryana) and Saharanpur district (Uttar Pradesh) fall in the category of "endemic" district.

1.5   JE Outbreaks in India

One of the most disturbing features of JE in India has been the regular occur-rence of outbreaks in different parts of the country. Almost every year, so-called "undiagnosed viral illness" invades India and unfailingly claims thousands of lives especially in children below 15 years. JE is claimed to be the leading cause in these episodes as it is the only disease causing viral encephalitis in India.

Kerala state in south India experienced its first outbreak in 1996, Haryana in 1990, and Uttaranchal, third state from Northern India, as recently as 2006 with 58 cases and 21 deaths. The worst ever recorded JE outbreak in India was reported from UP in 2005 when 6,061 cases with 1,500 deaths (CFR 24.7 %) were recorded from the state (Table 1.1). Significantly, higher CFR has been found during outbreaks, the highest reported being 84.8 % during 2005 outbreak in Haryana. Districts worst affected by outbreaks of Acute Encephalitis Syndrome (AES)/JE in terms of morbidity, mortality and frequency of occurrence are given in Table 1.2. Gorakhpur and Deoria districts of Uttar Pradesh have reported max-imum number of outbreaks so far. Although, primarily reported from rural agri-cultural areas, outbreaks have been reported from peri-urban areas also. The outbreaks are closely associated with monsoon, agricultural practice, presence of vector mosquito species and reservoir hosts.

1.6   Clinical Presentation and Diagnosis

Swelling of brain causes appearance of several signs and symptoms. Most com-mon being sudden onset of fever, abnormal movements, seizures (fits) and decrease in level of consciousness leading to unconsciousness in some cases.

Table 1.1 Situation of AES/JE in India during 2002–2007

Affected States/UT's	2002		2003		2004		2005		2006		2007 (P)	
	C	D	C	D	C	D	C	D	C	D	C	D
Andhra Pradesh	22	3	329	183	7	3	34	0	11	0	22	0
Assam	472	150	109	49	235	64	145	52	392	119	424	133
Bihar	8	1	6	2	85	20	192	64	21	3	336	164
Chandigarh	4	0	0	0	0	0	0	0	0	0	0	0
Delhi	1	0	12	5	17	0	6	0	1	0	0	0
Goa	11	0	0	0	0	0	4	0	0	0	61	0
Haryana	59	40	104	67	37	27	46	39	2	1	32	18
Karnataka	152	15	226	10	181	6	122	10	73	3	9	1
Kerala	0	0	17	2	9	1	1	0	3	3	0	0
Maharashtra	119	16	475	115	22	0	51	0	1	0	0	0
Manipur	2	1	1	0	0	0	1	0	0	0	65	0
Punjab	10	2	0	0	0	0	1	0	0	0	0	0
Tamil Nadu	0	0	163	36	88	9	51	11	18	1	37	0
Uttar Pradesh	604	133	1124	237	1030	228	6061	1500	2320	528	3024	645
West Bengal	301	105	2	1	3	1	12	6	0	0	16	2
Nagaland	0	0	0	0	0	0	0	0	0	0	7	0
Grand Total	1765	466	2568	707	1714	367	6727	1682	2842	658	4033	963

Source NVBDCP

Table 1.2 JE outbreaks, worst affected districts in India (1973–2006)

State/(Period)					
Andhra Pradesh (1979–2006)	Assam (1978–2006)	West Bengal (1973–2006)	Karnataka (1997–2006)	Uttar Pradesh (1978–2006)	Haryana (1990–2006)
Guntur	Dibrugarh	Burdwan	Kilar	Gorakhpur	Karnal
Anantapur	Lakhimpur	Bankura	Bellary	Deoria	Kurushetra
Chittoor		Birbhum		Saharanpur	
Cuddapah					
Prakasam					
Kurnool					

Case fatality rate generally varies from 20 to 40 %. Patients who recover may have neurological sequelae. These occur with variable frequency and depend on age and severity of illness. The commonly observed sequelae are mental impairment, severe neurological instability, personality changes and paralysis.

1.6.1 Laboratory Diagnosis

Diagnosis of Japanese encephalitis can be confirmed by isolation of virus from blood, CSF or brain biopsy and by serological tests. Isolation of virus is not done for diagnostic purposes. The serological tests include detection of IgM antibodies, which appear after the first week of onset of symptoms and are detectable for

Fig. 1.4 Transmission of JE
virus in nature

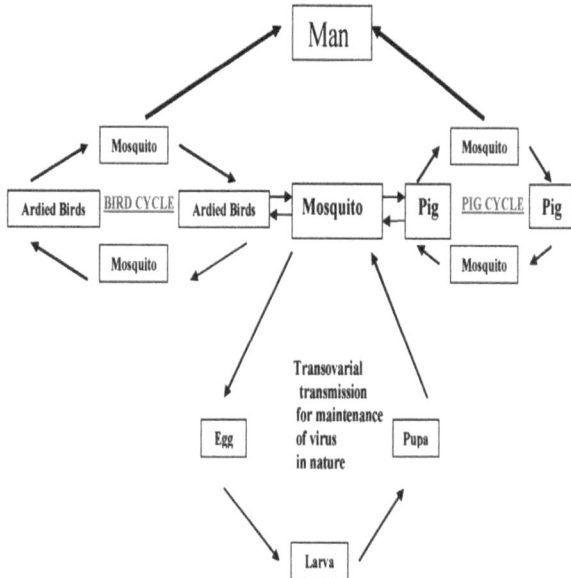

1–3 months after the acute episode. A 4-fold rise in IgG antibody tires in paired sets taken at an interval of 10 days is confirmatory. IgM antibodies indicate recent infection while IgG antibodies indicate previous infection.

   Detection of virus in vector mosquitoes is an important component of outbreak investigations to confirm the diagnosis of the disease in the outbreak. Japanese encephalitis (JE) virus-antigen was detected in unpreserved dry mosquitoes of *Culex tritaeniorhynchus* stored at room temperature for 30 days in south India (Tewari et al. 1999) and 20 months in Northern India (Das et al. 2005).

1.7   Epidemiology of Japanese Encephalitis

1.7.1   Natural History

JEV circulates and multiplies among pigs and birds. It is primarily an infection of animals and birds and the disease in human is incidental. The virus is transmitted through mosquitoes. In India, pigs and birds, particularly those belonging to Family Ardeidae, play important role in maintenance of JEV in nature (Fig. 1.4). *Ardeola grayii* (Pond heron) and *Bulbulus ibis* (Cattle egret) play a definite role in maintenance of JE in nature. Mosquitoes become infected by feeding on domestic pigs and wild birds infected with the JEV. Infected mosquitoes then transmit the JEV into humans and animals during the feeding process. Pigs are important amplifying host, they possess high body temperature (39 °C) and attract more vector mosquitoes. They develop tremendous viraemia but do not manifest any overt symptoms. The incidence of JE is recorded mainly among poorer people in villages where pig rearing is practiced. Cattle and buffaloes are preferred host for the mosquitoes, but they do not circulate the virus and therefore they do not play a role in its maintenance and spread, but act as "dampers" in the cycle. Man, is only occasionally infected and is a "dead end", since viraemia in human blood is too low and transient to infect mosquitoes. The disease occurs in human population when environmental conditions are favourable for prolific mosquito breeding as in monsoon and post monsoon season. The disease is mainly prevalent in rural areas among the lower socio-economic group of the population.

1.7.2 Seasonality

JE outbreaks are seasonal phenomena. A low-grade perennial transmission is recorded in areas with congenial climatic conditions throughout the year, e.g., southern India. Northern India receives summer monsoon and as such the transmission season begins from July to August, with incidence reaching peak in October to November, depending on the advancement of monsoon. With onset of winter, JE outbreak subsides.

Analysis of data of various JE outbreaks in the country reveals that vast majority of cases occurred in children. Though both sexes are affected, males usually outnumber females. In JE infection, inapparent infections outnumber the apparent infections. Estimated averages of only 1 in 300 JE viral infections result in symptomatic illness. The disease shows a scattered distribution with not more than 1–2 cases reported per village.

1.8 Vectors of Japanese Encephalitis

JEV has been isolated from ten species of the genus *Culex*, three species of *Anopheles* and three species of *Mansonia* in India (Phillip et al. 2000). Of these, the largest numbers of isolates have come from two species, *Culex tritaeniorhynchus* and *Cx. vishnui* which are incriminated as major vectors in Southern India. By large in whole of South East Asia and India, maximum number of JEV isolation have been done from members of Culex vishnui group of mosquitoes which include three species i.e., *Cx. tritaeniorhynchus* Giles, *Cx. vishnui* Theobald and *Cx. pseudovishnui* Colless. In addition to Cx. vishnui group other mosquitoes, viz. *Cx. bitaeniorhynchus* Giles, *Cx. gelidus* Theobald, *Cx. fuscocephala* Theobald, *Cx. infula* Theobald, *Cx. whitmorei* (Giles), *Cx. quinquefasciatus* Say, *Mansonia uniformis* (Theobald), *Mn. annulifera* (Theobald), *Mn. Indiana* Edwards, *Anopheles barbirostris* Van der Wulp, *An. peditaeniatus* (Leicester) and *An. subpictus* Grassi have also been incriminated/suspected as vector of JE.

The antigen capture enzyme immunoassay (ELA) was used as an improved JEV surveillance system by detecting JEV antigen from unpreserved dry vector mosquitoes stored at room temperature for 30 days (Tewari et al. 1999) and 20 months (Das et al. 2005).

In South India, the vectors of JEV mainly breed in paddy field water. At the start of the rainy season in July–August, mosquito density begins to rise with extensive paddy planting. This corresponds with the breeding season of ardeid birds in certain localised pockets. Virus transmission among pigs begins, and reaches its peak in September. Large numbers of infected mosquitoes have been detected at this time. The peak in human cases follows in October–November (Reuben and Gajanana 1997).

In the present study from Northern India, *Cx. tritaeniorhynchus* was found to be the predominant JE vector species and vertical transmission, in which the virus is transmitted from an infected female mosquito into her eggs appears to be the main mechanism of JEV transmission and persistence in nature.

1.9 Problem Faced While Investigating AES/JE Outbreak in Northern India

One of the most disturbing features of JE problem in India has been regular occurrence of outbreaks of AES/JE cases in different parts of the country (Table 1.1). In most of these outbreak investigations, adult mosquito survey indoors (usual protocol followed in Northern India) had shown negligible abundance of JE vector species in the area. This is in spite of the known fact that JE vectors predominantly rest outdoors among vegetation and only JEV is responsible for causing human encephalitis cases in India. As a result, there has been much confusion about the reason for the outbreak and the disease transmission could not be explained on the basis of inadequate entomological evidence.

The limited success in entomological investigation during repeated JE/AES outbreaks in different parts of Northern India is due to the following reasons:

1. *Lack of suitable, operationally feasible sampling tool/(s) available in the country* Almost every year, so-called "undiagnosed viral illness" invades India and unfailingly claims thousands of lives especially in children below 15 years. JE is claimed to be the leading cause in these episodes as it is the only disease causing viral encephalitis in India. Of these several hundred child deaths in India annually, >75 % are contributed by Northern India with case fatality ranging from 10 to 77.5 %.

   Health including control of mosquito vectors of human diseases is a state subject. In general, entomological man power and set-up available with the state authorities are either very poor or does not exist. Under such situation, it is not feasible to collect JE vectors during night hours using different types of baits as followed by workers having requisite facilities in institutes of Southern India.

2. *Correct identification of JE vector species is difficult* as the existing standard key to the species of *Culex* mosquitoes associated with JEV is difficult to

follow even by the trained entomologists available with the programme manager of the country.

3. *Limited knowledge on population dynamics including day resting sites of Cx. tritaeniorhynchus,* primary JE vector species operating in vast JE endemic areas of the country with the exception in Southern India from where population dynamics of local JE vector species have been fully demonstrated by experts of the local ICMR Institutes by incorporating highly skilled technologies and man power. Ecologically and culturally, there are lot of variations in Northern and Southern India. Only one rice crop per year is grown in Northern part while with extensive irrigation 2–3 rice crops per year is followed in many parts of South India. Extreme hot and cold climatic conditions exist in Northern part of the country, whereas in the southern part temperature, extremes are not so severe and a low-grade perennial transmission of the disease is recorded in areas with congenial climatic conditions throughout the year.

1.10 Research Problem Addressed by the Author and Included in this Book

1.10.1 Developed a New Sampling Technique for JE Vector Surveillance

In order to overcome the problem of inadequate entomological evidence during outbreak investigation of JE/AES, author of this book, developed "BPD hop cage method", a simple, cost-effective, operationally feasible surveillance tool specially designed to collect predominantly day resting adult JE vector species from land and aquatic vegetation (Das 2000; 2009). This surveillance tool has helped to study nearly every aspect of JE vector bionomics and establish entomological evidences of JE outbreaks occurring in the country upon its use since 2003 (Chap. 4 of this book).

1.10.2 Pictorial Key to Common Species of Culex (Culex) Mosquitoes Associated with JEV in India

Prepared a comparatively simple illustrated key for identification of mosquito vectors of EV in India and included in Chap. 3 of this book. Terminology used in this key is similar to an earlier key produced by the author for identification of Indian Anopheline mosquitoes (Das et al. 1990) which is in tune to the language used by majority of the users (entomologists, biologists and vector control programme managers) not only in India but also in other countries where the mosquito fauna is more or less similar.

1.10.3 Ecological Studies on JE Vectors in Northern India

This book mainly includes data generated by the author from ecological studies on JE vectors undertaken in two endemic and two non endemic areas of Northern India supported by data obtained during outbreak investigations of JE/AES carried out in different parts of the country by the central teams (constituted by the Ministry of Health and F. W., Government of India) in which the author was associated as a member (Chaps. 5, 6, 7).

1.10.4 Discovery of a New Bio-Control Agent

For the first time in science, a ciliated protozoa (*Chilodonella uncinata),* with no previous published record of pathogenicity against mosquitoes, was detected in non-endemic area of Sonipat District of Haryana state of India and was to be causing significant mortality in natural population of mosquitoes (*Cx. tritaeniorhynchus* and *Cx. pseudovishnui),* vectors of JEV (Das 2003). These are described in detail in Chap. 5: Sects. 5.6.5 and 5.7.2.

1.10.5 Situation Specific Vector Control Measures

Based on the observation obtained on JE vector abundance and JEV infection in local JE vector species in endemic and non endemic study areas of Northern India, situation specific vector control measures for local JE vectors were recommended to prevent transmission of the disease. Similar JE vector surveillance and vector control measures are recommended in other regions of South East Asia where similar ecological and environmental conditions exist.

1.11 JE Control Programme in India

In India, Japanese encephalitis control programme suffers from serious problem due to non-availability of cost effective control strategy. To contain the disease there are only two available intervention methods, viz. vaccination and vector control.

1.11.1 Vaccination

1.11.1.1 Global Scenario

Soon after the virus was discovered, crude vaccines were produced by the Japanese and others by growing the virus in mouse brain and inactivated in formalin. Production was refined and the vaccine efficacy demonstrated in Taiwan

(1960s) and Thailand (1980s). Vaccine developed a bad reputation in 1990s— (adverse reaction in 1/million). High production cost and the need for two or three doses plus booster and most importantly vaccination need to be completed 1 month before the transmission season in an area. However, rich countries like Japan controlled JE mainly through vaccination of human population and high standard of living.

1.11.1.2 Indian Experience

Indian JE vaccine is a Formalin inactivated vaccine made from the brain of suckling mice inoculated with the Nakayama JE strain, produced at the Central Research Institute (CRI), Kasauli, Himachal Pradesh, India. There have been some field trials in Gorakhpur district, during 1990s, the results were not encouraging.

Vaccination was thought to be not operationally feasible in India due to limited supply of vaccine to cover vulnerable age group who are at risk of JEV infection in vast endemic areas of the country (Murty et al. 2002). There is a stronger case for vaccinating children (high risk group) in high risk areas of our country, but the proposition is complicated by issues of: (i) Disease burden, (ii) Cost, and (iii) Competing health care priorities. In the year 2005, a massive outbreak of JE occurred in 34 districts of UP and adjoining districts of Bihar. In order to control JE, the Government of India introduced JE vaccination campaign in the 109 endemic districts in 15 states of the country in a phased manner over a period of 5 years 2006–2010 (Operational Guide, JE vaccination in India 2010). Under the programme, children between the age group of 1 and 15 years were vaccinated with a single dose of SA14-14-2 vaccine (made in China). Though there has been some impact reducing the case load and incidence of the disease in some districts, but it is not uniform especially in eastern UP.

1.11.2 JE Vector Control

Since *Cx. tritaeniorhynchus* mosquitoes, primary vector of JE in India, are predominantly exophilic in resting habit indoor residual spraying is not recommended to control the vector species. However, during the transmission season, malathion fogging in villages with a case is undertaken by the district health authorities to kill the infected mosquitoes to interrupt further transmission of the disease. In some endemic districts, anti-larval measures by Temephos @ 1.0 ppm in water bodies around the affected village are also being practiced. Still JE cases and deaths continue to occur. In view of the above, there is an urgent need to know the ecology of local JE vector followed by implementation of integrated JE vector control/management strategy.

References

Banerjee K (1988) Epidemiology of JE in India. In: Proceedings of the workshop on Japanese encephalitis, National Institute of Communicable Diseases, Delhi, pp 20–35, 18–22 January

Carey DE, Myres RM, Reuben R, Webb JKG (1969) Japanese encephalitis in South India: a summary of recent knowledge. J Indian Med Assoc 52:10

Chaterjee AK, Banerjee K (1975) Epidemiological studies on the encephalitis epidemic in Bankura. Indian J Med Sci 63:1164–1179

Das BP (2000) A new technique for sampling outdoor resting population of *Culex tritaeniorhynchus*, vector of Japanese encephalitis. In: Fourteenth national congress of parasitology, New Delhi, Abstr. No. PS-15, pp 133–134, 23–26 April 2000

Das BP (2003) *Chilodonella uncinata*—a protozoa pathogenic to mosquito larvae. Curr Sci 85:483–489

Das BP (2009) BPD hop cage method—a new device of collecting mosquitoes for effective JE vector surveillance. Invent Intell 44:24–25

Das BP, Rajagopal R, Akiyama J (1990) Pictorial key to the species of Indian Anopheline mosquitoes. J Pure Appl Zool 2:131–162

Das BP, Sharma SN, Kabilan L, Lal S et al (2005) First time detection of Japanese encephalitis virus antigen in dry and unpreserved *Culex tritaeniorhynchus* mosquitoes Giles, 1901, from Karnal district of Haryana state of India. J Commun Dis 37:131–133

Das Gupta A (1949) Economic and commercial geography. A. Mukherjee and Co. Ltd., Kolkata

Government of India (2010) Operational guide for Japanese encephalitis vaccination in India. Ministry of Health and Family Welfare, Government of India

Khare JB (1984) Current status of Japanese encephalitis in Uttar Pradesh. In: National conference on Japanese encephalitis held at New Delhi, Indian Council of Medical Research, New Delhi, pp 22–24, 3–4 November 1982

Mathur A, Chaturvedi UC, Tandon HO, Agarwal AK (1982) Japanese encephalitis epidemic in Uttar Pradesh, India during 1978. Indian J Med Res 75:161

Murty US, Satyakumar VR, Sriram K, Rao KM et al (2002) Seasonal prevalence of *Culex vishnui* subgroup, the major vector of Japanese encephalitis virus in an endemic district of Andhra Pradesh, India. J Am Mosq Control Assoc 18:290–293

Narasimham MVVL, Rao CR, Bendle MS, Yadava RL et al (1988) Epidemiological investigation of Japanese encephalitis outbreak in U.P. during 1988. J Commun Dis 20:253–275

Philip Samuel P, Hiriyan J, Gajanana A (2000) Japanese encephalitis virus infection in mosquitoes and its epidemiological implications. ICMR Bull 30:37–43

Reuben R, Gajanana A (1997) Japanese encephalitis in India. Indian J Paediatr 64:2433–2451

Sabesan S, Konuganti HKJ, Perumal V (2008) Spatial delimination, forecasting and control of Japanese encephalitis: India—a case study. Open Parasitol J 2:59–63

Sarkari NBS, Barthwal SP, Gupta AK, Singh MM et al (1984) A clinical appraisal of two epidemics of Japanese encephalitis in eastern Uttar Pradesh. In: National conference on Japanese encephalitis held at New Delhi, Indian Council of Medical Research, New Delhi, pp 34–40, 3–4 November 1982

Saxena VK, Baig MH, Bhardwaj M, Rajagopal R (1986) Entomological investigation of Japanese encephalitis outbreak Gorakhpur and Deoria Districts of Uttar Pradesh. J Commun Dis 18(3):219–221

Tewari SC, Thenmozhi V, Rajendran R, Appavoo NC et al (1999) Detection of Japanese encephalitis virus antigen in desiccated mosquitoes: an improved surveillance system. Trans R Soc Trop Med Hyg 93:525–526

Chapter 2
Mosquito Surveillance Tools Used and Methodology Followed in Ecological Study on JE Vectors in Northern India

Abstract A simple cost-effective, operationally feasible sampling method was developed for *Culex tritaeniorhynchus* predominantly exophilic JE vector mosquito species primarily for two reasons, viz: (i) to have an in-depth knowledge on the ecology of Mosquito vectors of Japanese encephalitis virus from Northern India, (ii) to overcome the problem regarding inadequate entomological evidence faced during repeated JE/AES outbreak investigation in different parts of India. The aim of this chapter is to describe the new mosquito surveillance tool used and methodology followed in ecological study on Mosquito vectors of Japanese encephalitis virus from endemic and non endemic areas of Northern India.

2.1 Introduction

Northern India comprising of National capital Delhi, Haryana and north western part of Uttar Pradesh (UP) constitute a transition semiarid zone between the arid Western India (Rajasthan) and humid belt of Eastern India (Bihar and West Bengal). Semiarid areas are characterised by high temperature (35–45 °C) during summer, hottest months (April–June) with precipitation less than potential evaporation. Almost all precipitation occurs as rains between July and September in these areas. Vegetation in semiarid zone is composed of stunted and scattered trees, shrubs, bushes and grasses.

2.2 Study Area

This study is based primarily on enormous field data accumulated by the author of this book (Bina Pani Das) on Japanese encephalitis (JE) vector bionomics from Northern India (Delhi, Haryana and Uttar Pradesh) at the national institute of communicable diseases (NICD), now known as national centre for disease control (NCDC), over a period of 9 years 1998–2006. Studies on JE vectors at NICD were broadly divided into three phases, viz.

B. P. Das, *Mosquito Vectors of Japanese Encephalitis Virus from Northern India*, SpringerBriefs in Animal Sciences, DOI: 10.1007/978-81-322-0861-7_2, © The Author(s) 2013

2.2.1 Phase 1: Non-Endemic Area—In and Around Delhi (1998–2002)

As the existing sampling tools were not adequate for predominantly exophilic JE vector (*Culex tritaeniorhynchus*) in the area, a need-based simple technique "Bina Pani Das (BPD) hop cage Method" was developed (Das 2000, 2009) to monitor vector abundance round the year. The tool developed and standardised at NICD was later found to be useful in studying nearly every aspect of adult JE vector bionomics (Chaps. 4, 5).

Studies on larval bionomics of JE vector in study areas of Haryana, both at field as well as laboratory, resulted into the discovery of a new bio-control agent (*Chilodonella uncinata*, a ciliate parasite) for mosquito vectors of human diseases (Malaria, Dengue and Chikungunya, JE)—(Das 2003, 2008). In view of potential biolarvicidal activity of *Ch. uncinata*, the Inventor and the Author of this book (Bina Pani Das) was allowed to file National and International patent applications during in 2001, for which entire financial assistance was provided by Department of Biotechnology, Ministry of Science and Technology (Das 2004). International search report was of "A" category and so far patent has been granted by six countries including USA (Patent # 7141245, date. 28.11.2006).

Initially *Ch. uncinata,* ciliate parasite was discovered in wild caught JE vector larvae (Das 2003). Thereafter, these were isolated, purified, colonised and when formulation was prepared it was found to be more effective against larvae of urban malaria vector (Das 2008).

2.2.2 Phase 2: Endemic Area I—Karnal District, Haryana State of India (2002–2004)

BPD hop cage Method was field tested and its efficacy was compared with those of conventional sampling tools (Drop net and Hand catch Method) for sampling outdoor resting population of vectors of Japanese encephalitis virus (JEV) in selected study villages of district Karnal (JE endemic), Haryana state of India during 2002–2004 (Chap. 6 of this book).

2.2.3 Phase 3: JE Endemic Area II—Saharanpur District, Uttar Pradesh State of India (2005–2006)

A longitudinal study, on ecology of vectors of JEV, was carried out in Saharanpur District, a JE endemic area of Uttar Pradesh, northern India to develop a spacio-temporal strategy for the control/management of JE vectors (Chap. 7).

2.3 Methodology

2.3.1 Metrological Data

Data on climatic conditions of Delhi during study period was collected from Mousam Bhawan, New Delhi and that of Karnal District (Haryana) and Saharanpur District (Uttar Pradesh) were collected from the Soil Research Institute, Karnal and Army establishment, Saharanpur, respectively.

2.3.2 Epidemiological Data

Data on suspected cases and deaths due to Acute Encephalitis Syndrom (AES)/JE during the study period and that of recent past of Karnal and Saharanpur District were collected from District Heath Officer, Karnal and Saharanpur District, respectively. While, data related to country as a whole from 1978 onwards were obtained from national vector-borne disease programme (NVBDCP).

2.3.3 Sampling of Adult Mosquitoes

Adult mosquitoes were collected and categorised as: (1) Outdoor collection; Indoor collections: (2) Hand catch, (3) Total catch by pyrethrum space spray and (4) Dusk collection.

2.3.3.1 Outdoor Collection by BPD Hop Cage Method

Outdoor resting mosquitoes were collected from 09:00 to 13:00 h from vegetations like jowar, mustard, and berseem at monthly interval from each study village using BPD hop cage method.

Construction hop cage The hop cage is a standard mosquito cage, measuring $30 \times 30 \times 30$ cm, each side made up of thin iron rod of 6 mm in diameter. The metal frame is covered with muslin cloth on its five sides, with a long cotton sleeve on the sixth side and a narrow cotton sleeve outlet on the opposite side. The sleeve of the cage is folded (Fig. 2.1) to allow entry of mosquitoes while hopping the cage on land and aquatic vegetation.

Mosquito collection method In order to collect resting mosquitoes from land vegetation, folded mosquito cage was allowed to hop through the shady vegetation near the ground (Fig. 2.2) and also low-level ground vegetation by a series of quick forward, backward, up and down movements through a distance of 5 ft (length wise) for about 2 min (Das 2000, 2009). This led to the trapping of mosquitoes

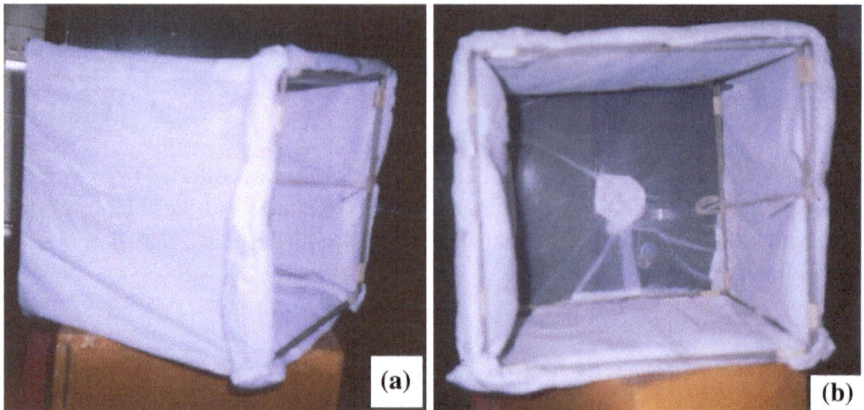

Fig. 2.1   Hop cage. **a** side view. **b** front view

Jowar plants   Green grasses   Hop cage

Fig. 2.2   Collection of mosquitoes resting in secondary ground vegetation following BPD hop cage method [Adapted from Das (2009)]

in the cage. One side open sleeve of the cage is then immediately folded in order to prevent the trapped mosquitoes from escaping. The mosquitoes were retrieved from the narrow cotton sleeve outlet of the cage by a mouth aspirator tube. Such attempts are made at least ten times in one type of vegetation. For collecting mosquitoes from aquatic vegetation like hyacinth marshes same procedure was followed along the side of the water body (Fig. 2.3).

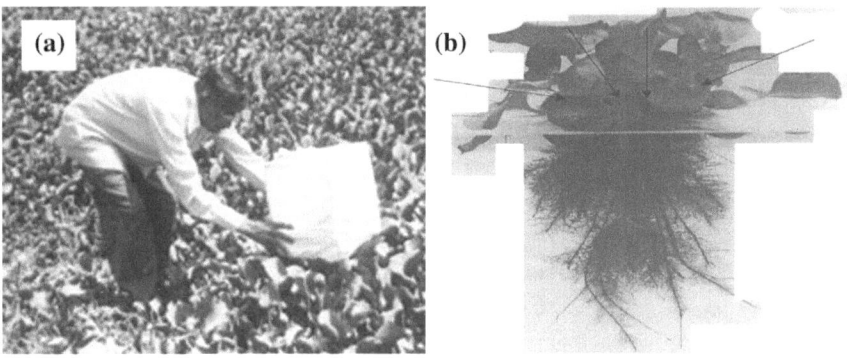

Fig. 2.3 a Collection of mosquitoes from aquatic vegetation (water hyacinth) using BPD hop cage method. **b** *Arrows* showing resting sites of mosquitoes among foliages of plant, just above the water level

Mosquito density measurement by hop cage Hop cage has to be used 10 times while collecting day-resting mosquitoes from land as well as aquatic vegetation and the mosquito density was measured as average number of female mosquitoes collected per ten hop cages (PTHC) by the following formula:

$$\text{Mosquito density (PTHC)} = \frac{\text{Total numbers of female mosquitoes collected}}{\text{Total numbers of hopping attempts made on vegetation}} \times 10$$

Each hop on vegetation covers an area of 1 sq. foot. The larger the area of the vegetation covered by hopping, the better representation of the mosquito density.

2.3.3.2 Indoor Collection

Indoor resting mosquitoes were collected from 06:00 to 08:00 h from human dwellings and cattle sheds in selected villages using Hand catch method spending 10 min per house. Mosquitoes were also collected during the day between 10:00 and 12:00 h from human dwelling (Total catch method) using pyrethrum space spray. Dusk collections were made around cattle shed and human dwelling by mouth aspirator tube and torchlight between 18:00 and 20:00 h. The abundance of mosquitoes in hand catch and in dusk collection was expressed as number of female mosquitoes collected per man hour (PMHD) and in Total catch as number of female mosquitoes collected per room density (PRD). Mosquito collections were transported to NICD/the respective field laboratories at Karnal and Saharanpur and sorted into sex, species using standard keys by Barraud (1934); Sirivanakarn (1976); Reuben et al. (1994), Das et al. (1990) and "Pictorial key to common species of *Culex* (*Culex*) mosquitoes associated with Japanese encephalitis virus in India"—Chap. 3 of this book. Abdominal condition of female mosquito species was noted and recorded.

2.3.4 Sampling of Mosquito Larvae

Larval collections were done by conventional dipping method. Fourth instar larvae were preserved using a simple standard method (Das 1986) and identified following Sirivanakarn (1976).

2.3.5 Mosquito Blood Meal Identification

Mosquitoes (*Culex tritaeniorhynchus, Cx. gelidus, Cx. bitaeniorhynchus, Cx. perplexus*) Leicester and *Aonopheles culicifacies* Giles collected from Saharanpur (July to October 2005) were processed at the National Institute of Malaria Research, Delhi for determining blood meal by enzyme linked immunosorbent assay (ELISA) following (Roy and Sharma 1987). The meal of fed mosquitoes was smeared on Whatman No. 3 filter paper by squeezing the stomach content of the mosquito, dried at room temperature and then kept in desiccator at 4 °C until use. For testing, the blood smears were eluted by cutting discs in microtitre plates containing 100 μl of 0.15 M Phosphate buffered saline (PBS), pH-7.2. The plates with eluates were kept at room temperature for 2 h followed by overnight at 4 °C. The elutes (~4 μl) were transferred by a clone-master template (Hyclone) from the wells onto a strip of nitrocellulose membrane (NCM) and dried at 37 °C for 10 min. After drying the NCM was washed thrice with 0.1 M Tris buffer (pH-7.5) containing 0.02 % Tween 20 (TBS-T). The NCM was then blocked with 1 % Bovine serum albumin (BSA) in TBS for 1 h at 37 °C. After blocking, the NCM was treated with anti-species globulin conjugated with peroxidase (Dakopatt, Denmark) at optimal dilution in PBS containing 1 % BSA. Finally, the reaction was visualised by enzyme specific substrate, 4-chloro-1-naphthol/H_2O_2 (SIGMA Aldrich). The reading of the coloured dots was denoted as $+/++/+++$ or $++++$ comparing with reference positive and negative samples.

2.3.6 JE Virus Detection

Culicine mosquitoes were sorted into pools, each containing about 50 specimens according to species, sex, place and date of collection. These pools were kept as dry specimens at room temperature (25 ± 2 °C). Mosquito pools were transported without cold chain facility to Centre for Research in Medical Entomology, Madurai, Tamil Nadu for detection of JEV infection in vector mosquito species. These were processed using an antigen-capture ELISA (Gajanana et al. 1995, 1997). For this assay, monoclonal antibody (MAB) 6B4A-10 reactive against JEV and a polyclonal antibody broadly reactive, conjugated with peroxidase, SLE MAB, 6B6C-1 were used as capture antibody and detector antibody, respectively. The ELISA plate contained known positive (JEV infected suckling mouse brain homogenate) and negative (homogenate of uninfected laboratory reared adult mosquito pools) controls. A mosquito pool was considered ELISA positive for virus antigen if its optical density (OD) value was greater than or equal to mean + standard deviation (SD) of the negative control (normal uninfected laboratory colony mosquitoes).

2.4 Pictorial Key for Identification of Common Species of *Culex (Culex)* Mosquitoes

A simplified key is prepared and incorporated in Chap. 3 of this book for identification of four medically important genera and 17 common species of culicine mosquitoes, ten of which have been incriminated as vectors of JEV in India. This key is used to identify culicine mosquitoes collected in the present work from the study areas of Delhi, Haryana, Uttar Pradesh, Uttaranchal, Chhattisgarh and Andhra Pradesh.

References

Barraud PJ (1934) The fauna of British India including Ceylon and Burma. Diptera. vol. V. Family Culicidae. Tribes Megarhinini and Culicini. Taylor and Francis, London

Das BP (1986) A simple modified method for mounting mosquito larvae. J Commun Dis 18:63–64

Das BP (2000) A new technique for sampling outdoor resting population of *Culex tritaeniorhynchus*, vector of Japanese encephalitis. In: Fourteenth national congress of parasitology, New Delhi, Abstr. No. PS-15, pp 133–134, 23–26 April 2000

Das BP (2003) *Chilodonella uncinata*—a protozoa pathogenic to mosquito larvae. Curr Sci 85:483–489

Das BP (2004) Process for preparation of a microbial control agent. Pub. No. US 2004/0219692, United States Patent Application Publication

Das BP (2008) New microbial insecticide—a discovery by accidental. Invent Intell 43:26–28

Das BP (2009) BPD hop cage method—a new device of collecting mosquitoes for effective JE vector surveillance. Invent Intell 44:24–25

Das BP, Rajagopal R, Akiyama J (1990) Pictorial key to the species of Indian Anopheline mosquitoes. J Pure Appl Zool 2:131–162

Gajanana A, Rajendran R, Thenmozhi V, Philip Samuel P et al (1995) Comparative evaluation of bioassay and ELISA for detection of Japanese encephalitis virus in field collected mosquitoes. Southeast Asian J Trop Med Public Health 26:91–97

Gajanana A, Rajendran R, Philip Samuel P, Thenmozhi V et al (1997) Japanese encephalitis in south Arcot district, Tamil Nadu, India: a three year longitudinal study of vector abundance and vector infection frequency. J Med Entomol 34:651–659

Reuben R, Tewari SC, Hiriyan J, Akiyama J (1994) Illustrated key to genera of *Culex (Culex)* associated with Japanese encephalitis in Southeast Asia ((Diptera: Culicidae). Mosq Syst 26:75–96

Roy A, Sharma VP (1987) Microdot ELISA: development of a sensitive and rapid test to identify the source of mosquito blood meals. Indian J Malariol 24:51–58

Sirivanakarn S (1976) Medical entomology studies III. A revision of the subgenus *Culex* in the Oriental Region (Diptera: Culicidae). Contrib Am Entomol Inst (Ann Arbor) 12(2):1–272

Chapter 3
Pictorial Key to Common Species of *Culex (Culex)* Mosquitoes Associated with Japanese Encephalitis Virus in India

Abstract The present work presents a key to 4 medically important genera and 17 common culicine species belonging to the subgenus *Culex (Culex)* associated with Japanese encephalitis in India. The key comprises 19 couplets (with their illustrations on facing pages), is easy to follow, since vivid illustrations accompany the sets of opposing characters in each couplet of the key. Language used in the key is tuned to the general users of the key (an advantage over the existing key in which difficult taxonomic terminology are used) not only in India but also in the neighbouring countries where the mosquito fauna is more or less similar.

3.1 Introduction

Mosquito family Culicidae is divided into three subfamilies, Anophelinae, Culicinae and Toxorhynchitinae with four medically important genera: *Anopheles, Culex, Aedes* and *Mansonia* (Table 3.1). In India, Japanese encephalitis virus has been isolated from ten species of the genus *Culex*, three species of genus *Anopheles* and three species of genus *Mansonia*. All the ten culicine vector species belong to the subgenus *Culex* Linnaeus, viz. *Cx. (Cx.) tritaeniorhynchus* Giles, *Cx. (Cx.) vishnui* Theobald, *Cx. (Cx.) pseudovishnui* Colless, *Cx. (Cx.) bitaeniorhynchus* Giles, *Cx. (Cx.) fuscocephala* Theobald, *Cx. (Cx.) infula* Theobald, *Cx. (Cx.) gelidus* Theobald, *Cx. (Cx.) epidesmus* (Theobald), *Cx. (Cx.) whitmorei* (Giles) and *Cx. (Cx.) quinquefasciatus* Say (Table 3.2). Of these, the largest numbers of JE virus isolates have come from two species, *Culex tritaeniorhynchus* and *Cx. vishnui* which are incriminated as major vectors.

Keys for identification of species belonging to the subgenus *Culex* mosquitoes have been published by earlier authors, the most important one is by Barraud's Fauna of British India volume on Culicidae, Tribes Megarhini and Culicini published in 1934. Even after 75 years, this work is still considered as a valuable reference work for taxonomists on culicine mosquitoes of Oriental region. Detailed description of each mosquito species with its known life stages was included in Barraud's work. He included 22 species in the subgenus *Culex* from undivided India, of which 20 are known from present-day India. Remaining two species: *nilgiricus* Edwards was later transferred to a different subgenus *Eumelanomyia*

B. P. Das, *Mosquito Vectors of Japanese Encephalitis Virus from Northern India*, SpringerBriefs in Animal Sciences, DOI: 10.1007/978-81-322-0861-7_3, © The Author(s) 2013

Table 3.1 Medically important genera[a] of mosquito family Culicidae (Diptera)

Subfamily 1.	Anophelinae		
		Genus 1.	*Anopheles*[a]
Subfamily 2.	Culicinae		
	Tribe 1. Aedeomyiini		
	Tribe 2. Aedini		
		Genus 2.	*Aedes*[a]
	Tribe 3. Culicini		
		Genus 3.	*Culex*[a]
	Tribe 4. Culisetini		
	Tribe 5. Ficalbiini		
	Tribe 6. Hodgesiini		
	Tribe 7. Mansoniini		
		Genus 4.	*Mansonia*[a]
	Tribe 8. Orthopodomyi-ini		
	Tribe 9. Sebithini		
	Tribe 10. Uranotaniini		
Subfamily 3.	Toxorhynchitinae		

a Medically important genus

by Sirivanakarn (1976) and *fusitarsis* Barraud 1924 was synonymized under *fuscocephala* by Bram (1967). F. W. Edwards (1911–1941), a well-known culicine mosquito taxonomist from Oriental, Palaearctic and Ethiopian Region also worked on culicine mosquitoes of undivided India and added an Appendix at the end of the book of Barraud's Fauna of British India (1934). *Cx. mimuloides* was originally recognised as a variety of *mimeticus* by Barraud (1924, 1934) but later, Edwards elevated it to specific rank among the members of his Mimeticus group.

After a gap of three decades of publication of Barraud's book, two valuable monographs have been published: a revision of the genus *Culex* in Thailand and a revision of the subgenus *Culex* in the Oriental Region by Sirivanakarn (1976). Both these publications are not easily available and being highly technical not practically feasible for field entomologists in India. *Cx. fuscifurcatus* Edwards was synonymised with *jacksoni* Edwards and *infula* Theobald was resurrected from synonymy with *bitaeniorhynchus* Giles by Sirivanakarn (1976).

Bram (1967) recognised two groups in subgenus Culex: Pipiens group and Sitiens group. He further chose to divide Sitiens group into four subgroups, viz. (i) Vishnui Subgroup, (ii) Gelidus Subgroup, (iii) Bitaeniorhynchus Subgroup and (iv) Sitiens Subgroup. Sirivanakarn (1976) recognised four subgroups in Pipiens group, viz. (i) Pipiens Subgroup, (ii) Trifilatus Subgroup, (iii) Theileri Subgroup and (iv) Univittatus Subgroup and added two more subgroup in Sitiens group, Barraud Subgroup and Mimeticus Subgroup. Indian fauna of culicine mosquitoes belonging to subgenus *Culex* comprises 25 species, of which six belong to the Pipiens group and 19 to Sitiens group. However, three species in Pipiens group are of restricted distribution mostly found in high altitudes while five species in

Table 3.2 Systematic list of Indian species of subgenus *Culex* Linnaeus

Pipiens group

1. *Cx. (Cx.) fuscocephala* Theobald, 1907
2. *Cx. (Cx.) hutchinsoni*[a] Barraud 1924
3. *Cx. (Cx.) quinquefasciatus* Say, 1823

4. *Cx. (Cx.) theileri*[a] Theobald, 1903
5. *Cx. (Cx.) univittatus* Theobald, 1901
6. *Cx. (Cx.) vagans*[a] Wiedemann, 1828

Sitiens group

7. *Cx. (Cx.) barraudi*[a] Edwards, 1922
8. *Cx. (Cx.) bitaeniorhynchus* Giles, 1901
9. *Cx. (Cx.) cornutus*[a] Edwards, 1922
10. *Cx. (Cx.) edwardsi*[a] Barraud, 1923
11. *Cx. (Cx.) epidesmus* (Theobald), 1910
12. *Cx. (Cx.) gelidus* Theobald, 1901
13. *Cx. (Cx.) infula* Theobald, 1901
14. *Cx. (Cx.) jacksoni*[a] Edwards, 1934
15. *Cx. (Cx.) mimeticus* Neo, 1899
16. *Cx. (Cx.) mimulus* Edwards, 1915

17. *Cx. (Cx.) mimuloides*[a] Barraud 1924
18. *Cx. (Cx.) perplexux* Leicester, 1908
19. *Cx. (Cx.) pseudovishnui* Colless, 1957
20. *Cx. (Cx.) sinensis* Theobald, 1903
21. *Cx. (Cx.) sitiens* Wiedmann, 1828
22. *Cx. (Cx.) tritaeniorhynchus* Giles, 1901
23. *Cx. (Cx.) vishnui* Theobald, 1901
24. *Cx. (Cx.) whitei* Barraud, 1923
25. *Cx. (Cx.) whitmorei* (Giles), 1904

Species Subgroup of Indian subgenus *Culex*

Pipiens group

Pipiens Subgroup. *quinquefasciatus*[b]
Theileri Subgroup. *theileri*
Trifilatus Subgroup. *vagans,*
hutchinsoni
Univittatus Subgroup. *fuscocephala*[b],
univittatus

Sitiens group

Barraudi Subgroup. *barraudi, edwardsi*
Bitaeniorhynchus Subgroup. *bitaeniorhynchus*[b],
cornutus, epidesmus[b], *infula*[b], *sinensis*
Gelidus Subgroup. *gelidus*[b]
Mimeticus Subgroup. *mimeticus, mimulus,*
mimuloides
Sitiens Subgroup. *sitiens, whitmorei*[b]
Vishnui Subgroup. *perplexux, pseudovishnui*[b],
tritaeniorhynchus[b], *vishnui*[b], *whitei*

[a]Uncommon species in India
[b]Incriminated as vectors of Japanese encephalitis virus in India

Sitiens group are not found in routine collection. The current taxonomic status of the species subgroup of subgenus *Culex* is shown in Table 3.2.

The members of the Culex Vishnui Subgroup of the subgenus *Culex* are very difficult to identify but entomologist and biologist engaged in vector control programme must be able to identify the local JE vector species correctly for effective delivery of vector control measures. In order to overcome the difficulty in identification of JE vectors, Reuben et al. (1994) published illustrated keys to the Indian genera of mosquitoes, the subgenera of *Culex* and the common Indian species of *Culex* mosquitoes associated with Japanese encephalitis. However, identification of Indian species of *Culex* mosquitoes involved in transmission of JE virus continues to be difficult for the peripheral public health field workers compelled me to develop the present key to common species of *Culex* mosquitoes associated with Japanese encephalitis in India in the pattern of pictorial key that was developed earlier by us (Das et al. 1990) for the species of Indian Anopheline mosquitoes.

Table 3.3 Terminology used in mosquito taxonomy

Barraud (1934)	Sirivanakarn (1976); Reuben et al. (1994)
Thorax	
mesonotum	Scutum
Wing	
Vein 1	radius—one (R_1)
Vein 2	radius—two-plus-three (R_{2+3})
Vein 2.1	radius—two (R_2)
Vein 2.2	radius—two (R_3)
Vein 3	radius—four-plus-five (R_{4+5})
Vein 4	media—one-plus-two (M_{1+2})
Vein 4.1	media—one (M_1)
Vein 4.2	media—two (M_2)
Vein 5	cubitus—(Cu)
Vein 5.1	cubitus—one (Cu_1)
Vein 5.2	cubitus—two (Cu_2)
Vein 6	anal vein

The present work provides a key to the four medically important genera of the mosquito family culicidae and 17 common culicine species belonging to the subgenus *Culex* associated with Japanese encephalitis in India. Since a few species (*mimeticus*, *mimulus*, Vishnui subgroup) can be differentiated by larval and adult characters, the key is not based only on adult female characters. Mosquito larval mounts are easily prepared following a very simple method Das (1986). In the present key, the differentiating characters in each couplet and their illustrations are on facing pages as followed elsewhere (Das et al. 1990), so as to facilitate identification. Numbers and letters in brackets in the couplets refer to the particular structure in the illustrations that will help the user. Important morphological characters of adults and immature stages of culicine mosquitoes used for identification of various species are given in Figs. 3.1 and 3.2. I have followed the terminology used by Barraud (1934) as bulk of would be users (field entomologists, biologist and peripheral public health field workers) in India are familiar with this terminology. A comparison of the more appropriate terminology used by Sirivanakarn (1976); Reuben et al. (1994) is shown in Table 3.3. No originality is claimed for the key presented, as it is based on descriptions of Barraud, Sirivanakarn, Bram, Reuben et al., Das and Kaul and Das et al.

While identifying specimens from a new area particularly in Northern India one has to be very careful in separating apparently similar species under similar situation. In Karnal (Endemic district of Haryana state of India) during JE transmission season, large number of *Cx. perplexus*, a non-vector species was collected along with *Cx. tritaeniorhynchus,* vector species in the area. These species can be easily identified using the present key. However, in case of any doubt the user of this key is advised to refer to Sirivanakarn (1976) to avoid misidentification.

PICTORIAL KEY

Morphology of *Culex*: ADULT

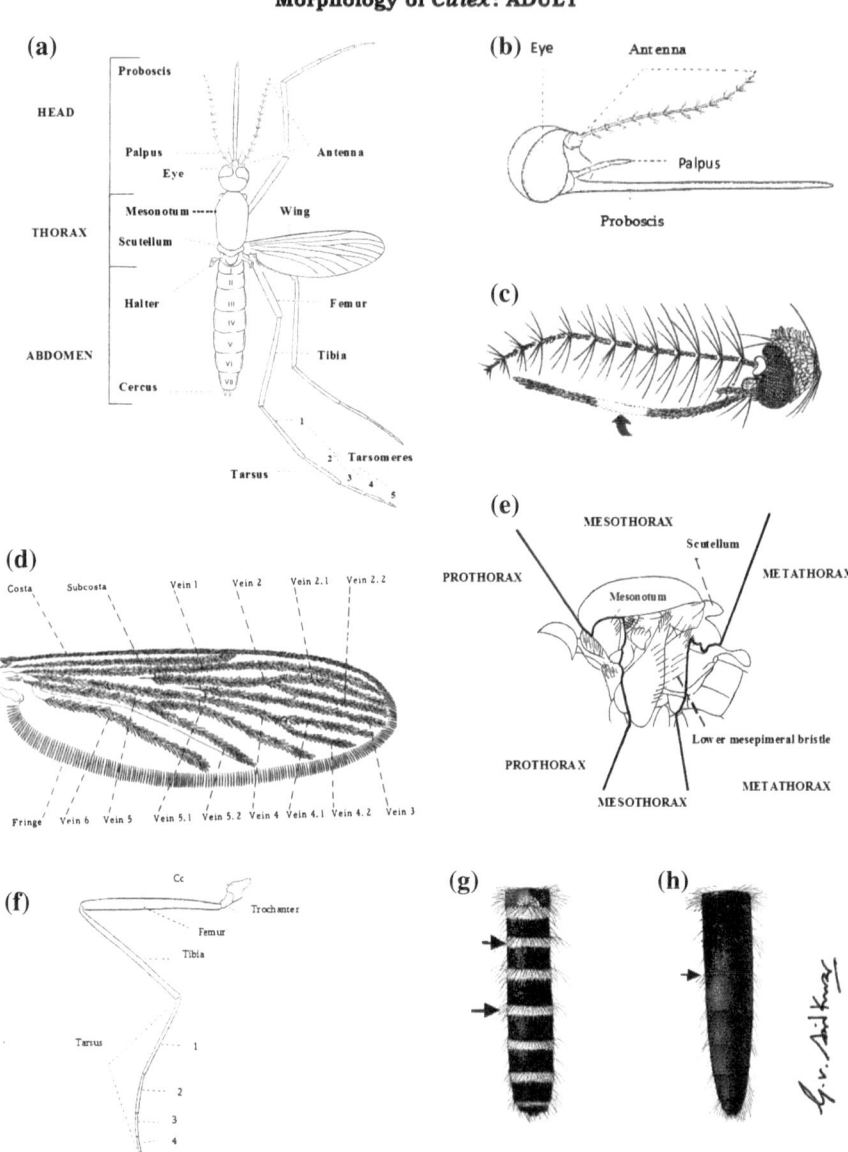

Fig. 3.1 a Adult, dorsal view. **b** Lateral view of head. **c** Proboscis with median pale band. **d** Wing venation. **e** Thorax, lateral view. **f** Leg. **g** Abdomen with basal pale band. **h** Abdomen without any pale band (Partly adapted from Richard et al. 1990; Pratt and Stojanovich 1966)

Morphology of *Culex*: LARVA

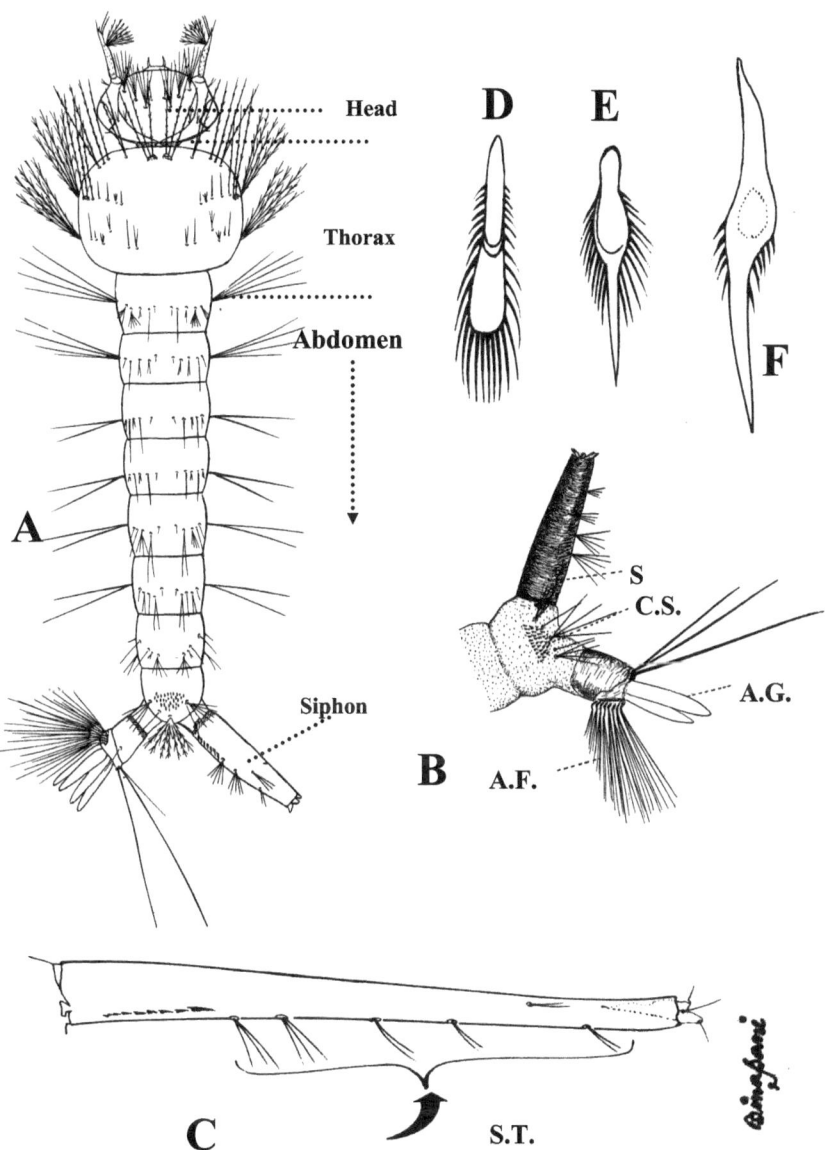

Fig. 3.2 a Full grown Larva, dorsal view. **b** Terminal segments of a fourth stage larva: *A.F.*, anal fin; *A.G.*, anal gill; *C.S.* comb scale; *S*, syphon. **c** Syphon, lateral view: *S.T.*, syphonal tuft. **d** comb scale, fan shaped. **e, f** comb scales with strong median tooth (Partly adapted from Pratt and Stojanovich 1966a, b; Richard et al. 1990)

3.2 Pictorial Key to Commonly Encountered Species of *Culex* (*Culex*) Mosquitoes Associated with Japanese Encephalitis in India

(Unless otherwise stated the characters used in the key generally refers to adults of both species. However, in males: palpi are longer than proboscis and pointed)

1   Palpi [a] as long as proboscis (b)   *Anopheles* species

         Palpi [a] much shorter than proboscis (b)   2

2 (1) Wings with broad asymmetrical scales, pale and dark scales intermixed; wings distinctly speckled

                                                                   ***Mansonia* species**

         Wing scales never broad or asymmetrical, speckling if present indistinct may have spots of white scales as in many anophelines

                                                    ***Aedes, Culex* species**   3

3 (2) Legs with white bands (a); abdomen pointed (b)

                                                                     ***Aedes* species**

         Legs unbanded, entirely dark or terminal segments white (a); abdomen blunt (b)

                                                    ***Culex* species**   4

4 (3) Proboscis and tarsi with pale rings (a); lower mesepimeral bristle absent (b)   7

         Proboscis (a) and tarsi (b) without any pale ring; 1 or 2 lower mesepimeral bristle present (c)   5

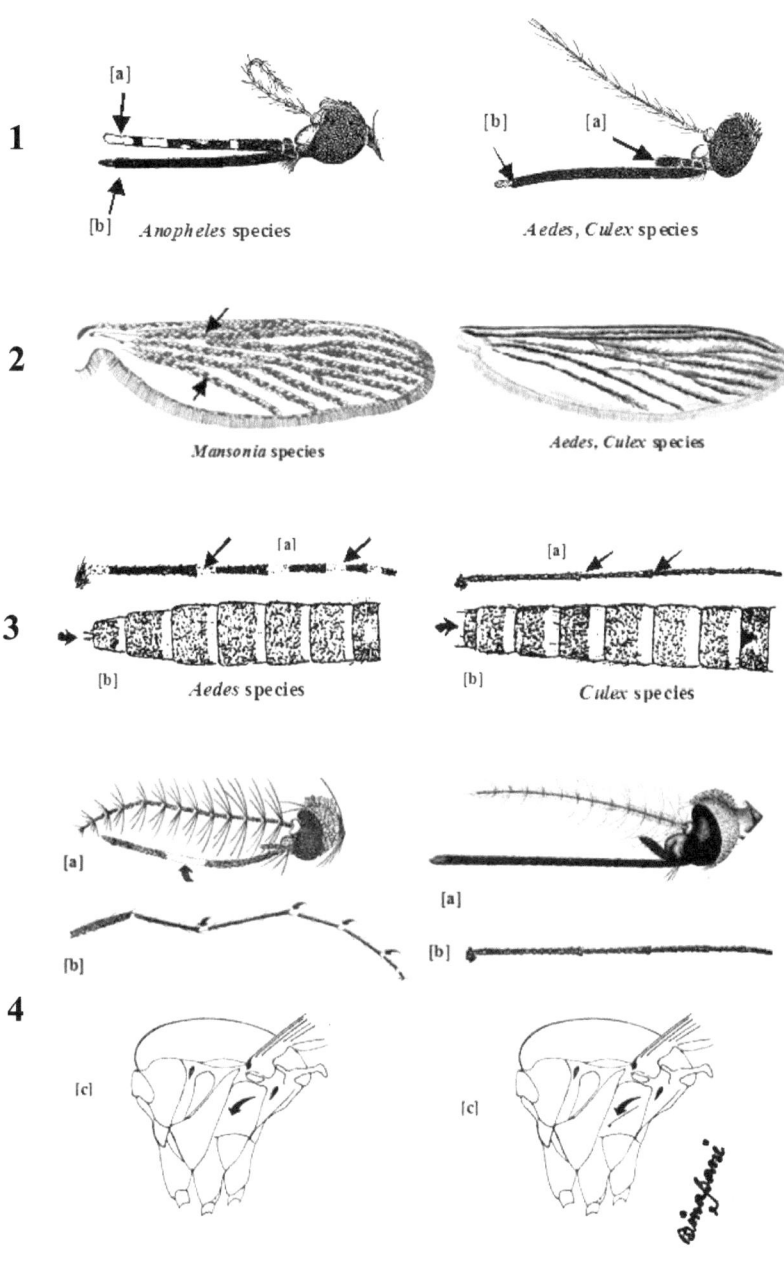

1   *Anopheles* species   *Aedes, Culex* species

2   *Mansonia* species   *Aedes, Culex* species

3   *Aedes* species   *Culex* species

4

5 (4) Abdominal terga with basal transverse pale bands (a); pleuron without distinct pattern of dark and pale stripes (b)   6

Abdominal terga without basal transverse pale bands (a); pleuron with distinct pattern of dark and pale stripes (b)

fuscocephala

6 (5) Postspiracular scales present (a); abdominal terga II–VI with even basal pale bands

univittatus

Postspiracular scales absent (a); terga II–VI with evenly broad or slightly medially produced basal pale bands

quinquefasciatus

7 (4) Wings with pale spots or yellow patches or speckled with light or dark scales   **8**

Wings dark scaled   **12**

8 (7) Wings with pattern of pale spots resemble those of anophelines   **9**

Wings either with light and dark scales or with yellow patches   **10**

9 (8) Vein 3 with pale area in the middle (a); first costal pale spot of wing involves costa and subcosta (b). Larva: individual comb scale with apical fringe terminating in a strong, distinct median spine (c)

mimeticus

Vein 3 entirely dark (a); first costal pale spot of wing involves costa, subcosta and vein 1 (b). Larva: individual comb scale with apical fringe terminating in a weak median spine (c)

mimulus

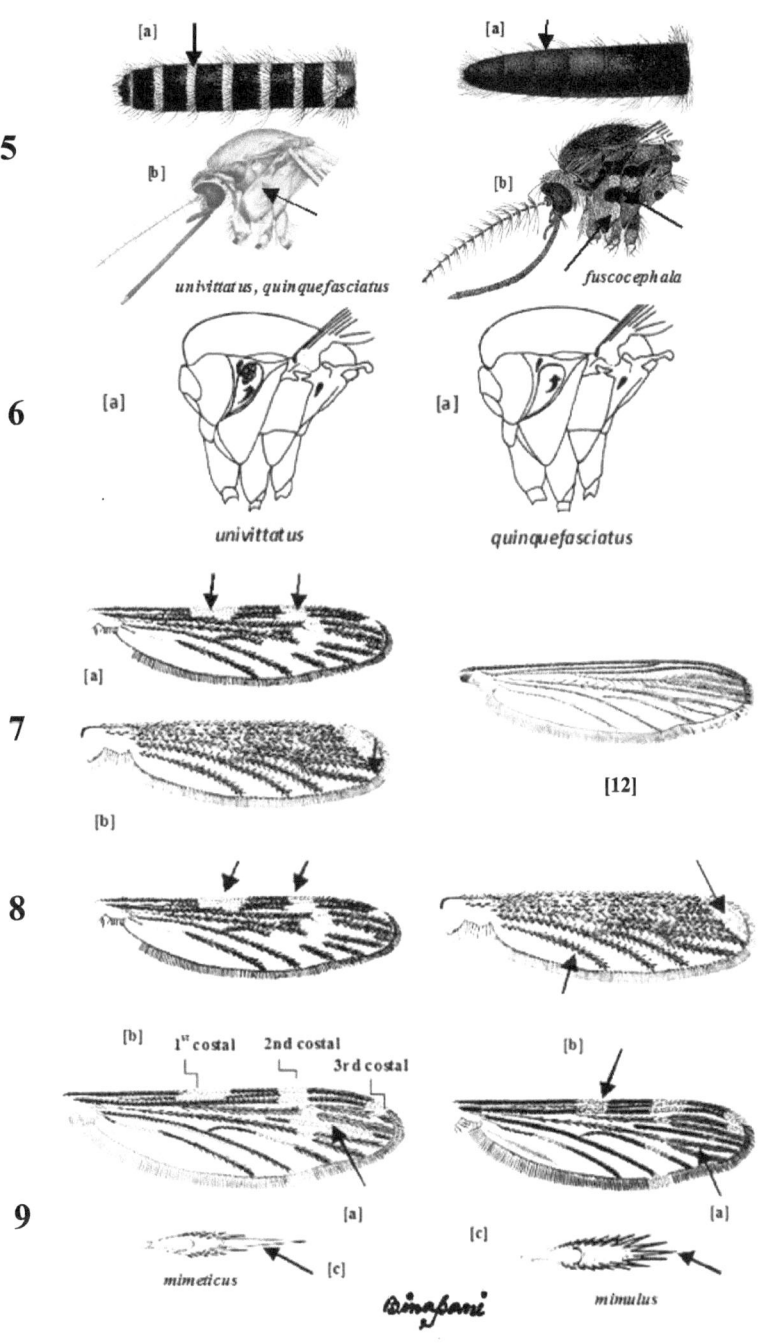

5   [a]   [b]   *univittatus, quinquefasciatus*   [a]   [b]   *fuscocephala*

6   [a]   *univittatus*   [a]   *quinquefasciatus*

7   [a]   [b]   [12]

8

9   [b]   1st costal   2nd costal   3rd costal   [a]   [c]   *mimeticus*   [b]   [a]   [c]   *mimulus*

10 (8)   Abdominal terga II–VIII mainly yellowish or golden (a); wings with a yellowish patch near tip of vein 1, 2.1, 2.2 and 3 (b); proboscis with median yellow band occupying nearly two-thirds of its length

epidesmus

Abdominal terga II–VIII with dark and pale bands (a); wing tip without definite pale area; wings heavily or moderately speckled with pale scales (b)   **11**

11 (10)   Abdominal terga II–VIII with evenly broad apical pale bands and without apicolateral pale patches (a); dark scaled areas of all abdominal terga heavily sprinkled with pale scales (b); legs and wings heavily speckled with pale scales (c)

bitaeniorhynchus

Abdominal terga II–IV largely dark (a), or with narrow apical bands, apicolateral yellowish patches and median basal pale bands or spots; V–VII with apical and basal bands, basal bands narrower (b); legs and wings moderately speckled (c)

infula

12 (7)   Basal two-thirds of mesonotum distinctly whitish or pale greyish, apical one-third dark   **13**

Basal two-thirds of mesonotum not distinctly paler than the apical one-third   **15**

13 (12)   Abdominal terga II–VIII with narrow more or less even transverse apical and basal bands, apical bands (a) usually broader and more distinct than basal bands (b)

sinensis

Abdominal terga with basal bands only   **14**

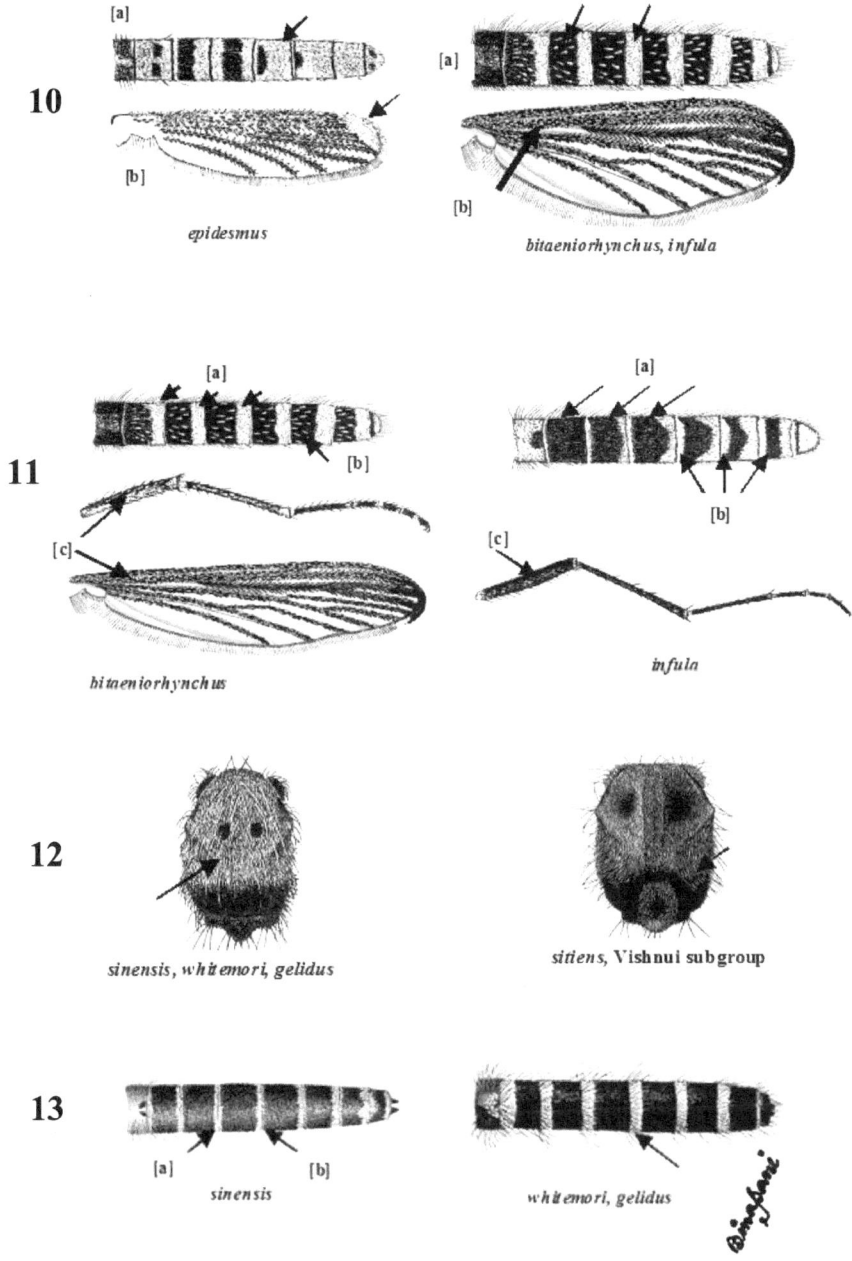

10

epidesmus

bitaeniorhynchus, infula

11

bitaeniorhynchus

infula

12

sinensis, whitemori, gelidus

sitiens, **Vishnui subgroup**

13

sinensis

whitemori, gelidus

14 (13) Abdominal terga II–VII with triangular basal pale patches (a); pale band on proboscis broad, as long as or longer than basal dark area (b); anterior surface of fore and hind femora speckled with pale scales (c)

whitemorei

Abdominal terga II–VI with complete basal pale bands producing strongly in the middle, terga VII–VIII with narrow even basal pale bands (a); pale band on proboscis narrow, much shorter than basal dark area (b); anterior surface of fore and mid femora not speckled with pale scales (c)

gelidus

15 (12) Wing scales mainly dark (a); scutal integument dark (b); hind tarsus dark, with basal pale band on tarsomeres I–IV (c)

sitiens

Wings with few to scattered pale scales (a); scutal integument light brown (b); tarsomeres I–IV with or without narrow, poorly defined apical and basal pale band

Vishnui Subgroup   **16**

16 (15) Proboscis usually with accessory pale patches or pale spots (speckling) on ventral surface (a); hind femur pale with an apical dark ring (b); thoracic scales dark coppery gold. Larva: individual comb scales apically rounded, fringed with subequal spicules (c); syphon (7–8)x as long as wide, syphonal tufts weak, with 2–5 branches (d)

tritaeniorhynchus

Proboscis without accessory pale patches or pale spots (a); hind femur without apical dark ring but with dark and pale areas, with or without speckling. Larva: individual comb scales with a weak or well-developed strong median apical spine and short subequal fringe like lateral spines (c); syphonal tufts variable   **17**

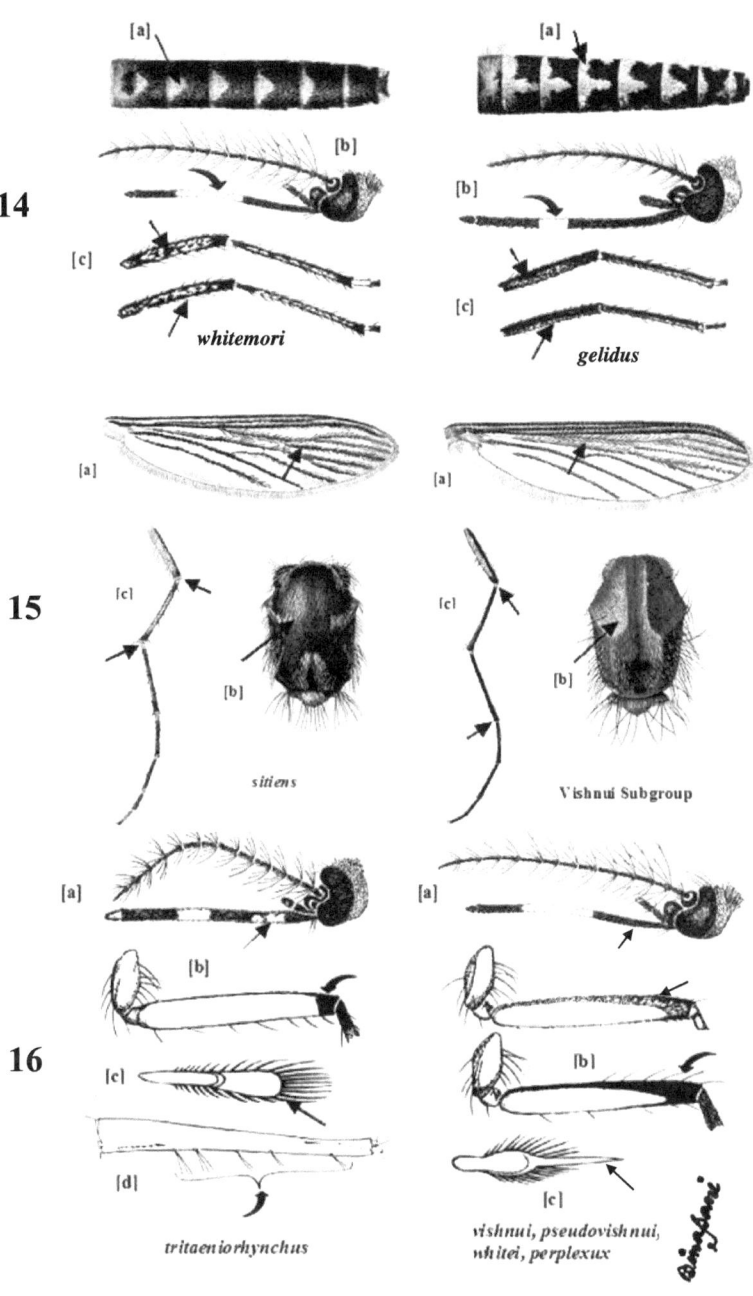

14

whitemori

gelidus

15

sitiens

Vishnui Subgroup

16

tritaeniorhynchus

*vishnui, pseudovishnui,
whitei, perplexux*

17 (16)   Larva: syphon straight, syphonal tufts usually bifid, weak, short, as long as or less than the syphonal width at their point of attachment (a); comb scales numerous, about 40, all small, subequal in size, individual scale with a weak median terminal spine (b). Abdominal terga II–VII with or without narrow basal pale band

whitei

Larva: syphon straight or lightly to strongly curved apically, syphonal tufts long, strong, with four or more branches (a)   **18**

18 (17)   Larva: Syphon lightly curved, syphonal tufts strong, with six or more branches (a); about 22 comb scales similar in size and length, arranged in a broad oval patch, individual scale with apical fringe terminating into a strong median spine (b). Hind femur with pale and dark areas poorly contrasted (c); thoracic scales brown to pale straw

vishnui

Larva: Syphon lightly curved (a) or straight (b); comb scales 4–20, arranged in 1 or 2–3 irregular rows   **19**

19 (18)   Larva: Syphon lightly curved (a); comb scales 5–7, large, strong, spiniform (b), arranged in a single row. Hind femur with dark band apically and dorsally, usually well contrasting with pale areas (c); thoracic scales yellow-gold to silvery white.

pseudovishnui

Larva: Syphon slender, straight (a); comb scales 12–20, short, with strong apical spine (b) arranged in 2–3 irregular rows. Abdominal terga II–VII entirely dark or with narrow basal pale band; integument of scutum dark brown to blackish

perplexux

Some of the illustrations in this key have been adapted from Sirivanakarn (1976), Bram (1967), Reuben et al. (1994), Das et al. (1990) and Das and Kaul (1998).

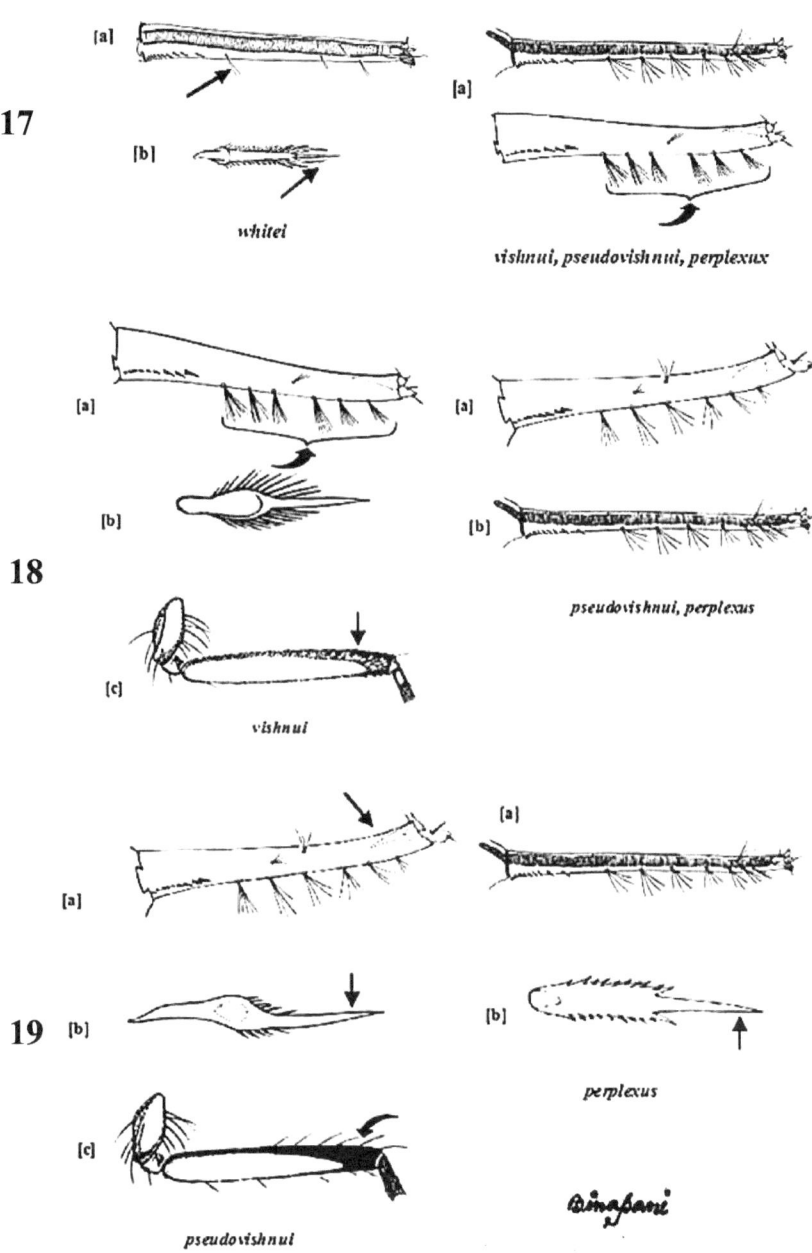

17

[a]

[b]

whitei

[a]

vishnui, pseudovishnui, perplexus

18

[a]

[b]

[c]

vishnui

[a]

[b]

pseudovishnui, perplexus

19

[a]

[b]

[c]

pseudovishnui

[a]

[b]

perplexus

References

Barraud PJ (1924) A revision of culicine mosquitoes of India. Part XI. Some Indian species of *Culex* L. Indian J Med Res 11:979–998

Barraud PJ (1934) The fauna of British India including Ceylon and Burma. Diptera. Vol. V. Family Culicidae. Tribes Megarhinini and Culicini. Taylor and Francis, London

Bram RA (1967) Contributions to the mosquito fauna of Southeast Asia—II. The genus *Culex* in Thailand (Diptera: Culicidae). Contrib Am Entomol Inst (Ann Arbor) 2(1):1–296

Das BP (1986) A simple modified method for mounting mosquito larvae. J Commun Dis 18:63–64

Das BP, Kaul SM (1998) Pictorial key to the common Indian species of Aedes (Stegomyia) mosquitoes. J Commun Dis 30:123–127

Das BP, Rajagopal R, Akiyama J (1990) Pictorial key to the species of Indian Anopheline mosquitoes. J Pure Appl Zool 2:131–162

Pratt HD, Stojanovich CJ (1966) Workbook on the identification of *Anopheles* Adults. U.S. Department of Health, Education, and Welfare, Public Health Service, Atlanta (Gorgia, U.S.A.), pp 56

Pratt HD, Stojanovich CJ (1966) Workbook on the identification of *Anopheles* Larvae. U.S. Department of Health, Education, and Welfare, Public Health Service, Atlanta (Gorgia, U.S.A.), pp 63

Reuben R, Tewari SC, Hiriyan J, Akiyama J (1994) Illustrated key to genera of Culex (*Culex)* associated with Japanese encephalitis in Southeast Asia ((Diptera: Culicidae). Mosq Syst 26:75–96

Richard F, Darsie J, Shreedhar PP (1990) The mosquitoes of Nepal: their identification, distribution and biology. Mosq Syst 22(2):69–130

Sirivanakarn S (1976) Medical Entomology studies III. A revision of the subgenus *Culex* in the Oriental Region (Diptera: Culicidae). Contrib Am Entomol Inst (Ann Arbor) 12(2):1–272

Chapter 4
BPD Hop Cage Method for Effective JE Vector Surveillance

Abstract Accidental detection of natural day resting sites of *Culex tritaeniorhynchus*, primary vector of Japanese encephalitis, among aquatic vegetation in East Delhi was the inspiration behind the development of a new sampling technique "BPD hop cage method" as the conventional tools were found to be insufficient to collect mosquitoes resting in aquatic vegetation. The new sampling technique was found to be equally effective in monitoring the abundance of *Cx. tritaeniorhynchus* mosquitoes resting in land vegetation. This chapter presents the different phases in standardisation of BPD hop cage method in study areas of Delhi and Haryana and its use in outbreak investigation of JE/AES as well as JE vector surveillance round the year particularly in Northern India which contributes over 75 % of incidence of Japanese encephalitis cases from India.

4.1 Introduction

Vector surveillance is used to determine changes in the geographical distribution and density/abundance of the vector species to, detect/isolate disease-causing agent in vector mosquitoes, evaluate vector-control programmes, determine relative measurements of the vector population over time and to facilitate appropriate and timely decisions regarding interventions. Vector abundance and vector infection are the two important parameters for vector surveillance (Philip Samuel et al. 2000). *Culex tritaeniorhynchus Giles* is a recognised major vector of Japanese encephalitis in India based on the fact that: (i) this species has been the most prevalent species reported from JE outbreak areas in northern and southern India (Rajagopalan et al. 1978; Kaliwal 1991; Mathur et al. 1981; Das et al. 2004) and (ii) repeated isolation of JE virus (JEV) from this species from many parts of India (Carey et al. 1968; Murthy et al. 2000; Arunachalam et al. 2002; Das et al. 2005).

 Cx. tritaeniorhynchus is mainly exophilic in its resting behaviour (Reuben 1971) and has been reported to be absent or represented with very low densities from indoor situation during JE outbreaks (Das et al. 2004; Saxena et al. 1986; Sharma et al. 1991). Adults of these mosquitoes rest among various types of wild and cultivated vegetation in varied type of agroecological areas of the country. This type of resting habitats was not fully explored for monitoring vector abundance

B. P. Das, *Mosquito Vectors of Japanese Encephalitis Virus from Northern India*, SpringerBriefs in Animal Sciences, DOI: 10.1007/978-81-322-0861-7_4, © The Author(s) 2013

or information available based on only one-time study (Kulkarni and Rajput 1988; Khan et al. 1997). Monitoring the adult vector population, as a prerequisite for instituting control measures, has been a challenge to the vector control programme managers. In the programme, the usual method employed to monitor abundance of adult mosquitoes in India is by the use of aspirator tube following hand catch method from indoor situation although this method is not adequate for a predominantly exophilic species like *Cx. tritaeniorhynchus*, primary vector of Japanese encephalitis in India. Direct collection of resting mosquitoes from natural vegetation was primarily recommended for sampling mosquitoes from outdoor habitats, following hand catch and drop net method from outdoor resting sites in vegetation (World Health Organization 1975). However, these methods have limitations: the hand catch method is time consuming and unrewarding, the drop net is too large to be carried and requires three workers for field operation (Das 2009).

During May 1999 with Temperature >40 °C, in Osmanpur village, East Delhi, while collecting mosquito larvae, freshly fed and semigravid population of *Cx. tritaeniorhynchus* adults were found resting underneath the leaves as well as foliages of water hyacinth where drop net cannot be used. With lot of theoretical and practical exercises in the laboratory as well as in the field for about 3–4 months emerged the new method which was initially designated as "Sweep Cage Method" (Das et al. 2004; Gupta et al. 2005). This method since involves hopping procedure in the field, I have renamed it as Bina Pani Das (BPD) hop cage method (Das 2009). The new methodology was found to be equally effective in sampling mosquitoes resting on land vegetation (Fig. 2.2, explained in Chap. 2). Thereafter, in order to include the new sampling tool (BPD hop cage method) in JE vector surveillance, follow-up studies were undertaken in selected areas of Delhi and Haryana. These are:

4.2 Standardisation of BPD Hop Cage Method for Sampling JE Vectors

Study area: In order to standardise BPD hop cage method, vector surveillance was carried out to know the JE vector densities in two villages in and adjoining areas of Delhi, viz. Madanpur Khadar, a non-paddy cultivated area in south Delhi and Safiabad (Sonipat district, Haryana) where paddy is cultivated extensively. These two villages are located 80-km apart from each other and were selected as there were various types of land and aquatic vegetation almost round the year. The study was conducted once in a month from September to December 1999, wherein adult mosquitoes were collected from both indoor and outdoor situations to know the JE vector abundance in the area.

Madanpur Khadar (Map 4.1) is situated on the bank of Agra canal in South Zone, Delhi. There is an irrigation channel and a few ground-water pools supporting prolific growth of water hyacinth (*Eichhornia crassipes*) and elephant grass. Land vegetation in the area includes: plenty of wild grasses and seasonal vegetables

Map. 4.1 Study sites of
Delhi (Madanpur Khadar,
Osmanpur) and Haryana
(Safiabad)

like cauliflower (*Brassica oleracea var. botrytis*), cabbage (*Brassica oleracea var. capitata*), raddish (*Raphanus sativus*), brinjal (*Solanum melongena*), etc.

Safiabad village is situated in Haryana adjacent to Narela Zone, close to Singhu border, north of Delhi. Paddy is cultivated extensively during July–October in the area. Millet (*Sorghum bicolor*) is also cultivated almost simultaneously, which support ground vegetation (green grasses) for a fairly long period. Other land vegetation includes wheat (*Triticum aestivum*), berseem, a fodder plant (*Trifolium alexandrinum*), seasonal crops: carrot (*Daucus carota*), marigold (*Calendula officinalis*), brinjal, cauliflower, cabbage, spinach (Basella alba), etc., and a few species of wild plants including hemp. There are a few ponds and roadside burrow pits on the outer side of the village but unlike village Madanpur Khadar, water hyacinth is totally absent from the entire area of Safiabad village.

Indoor collections Indoor resting collections were made during 06.00 to 08.00 h in cattle sheds by insect collectors using standard aspirator tubes and flash lights and the density measured as number of female mosquitoes collected per man hour (PMH).

Outdoor collections Outdoor resting mosquitoes were collected during 09.00 to 13.00 h from land and aquatic vegetations following BPD hop cage method (as explained in Chap. 2) and the density of the mosquito species is calculated as number of female mosquitoes collected Per Ten Hop Cages (PTHC) by the following formula:

$$\text{Mosquito density (PTHC)} = \frac{\text{Total numbers of female mosquitoes collected}}{\text{Total numbers of hopping attempts made on vegetation}} \times 10$$

Collections were immediately transported to Vector Ecology Laboratory, National Institute of Communicable Diseases (NICD), sorted according to species, identified based on standard keys (Sirivanakarn 1976; Reuben et al. 1994; "Pictorial key to common species of Culex (Culex) mosquitoes associated with Japanese encephalitis virus in India"—Chap. 3 of this book; Das et al. 1990) and female abdominal condition (unfed, freshly fed, semigravid and gravid) was recorded.

During a period 4 months (September–December, 1999), a total of 275 and 192 mosquitoes were collected from outdoor situation, resting in grassy ground vegetation of Jowar field and water hyacinth marshes, and 125 and 366 mosquitoes from indoor situation from Safiabad and Madanpur Khadar, respectively. Outdoor collections revealed four species of mosquitoes *Cx. tritaeniorhynchus*, *Cx. quinquefasciatus* Say, *Armigeres subalbatus (Coquillett)* and *Mansonia annulifera (Theobald)*, while indoor collection revealed eleven species: *Cx. tritaeniorhynchus*, *Cx. fuscocephala Theobald*, *Cx. pseudovishnui Colless*, *Cx. quinquefasciatus*, *Cx. whitei Barraud*, *Anopheles annularis van der Walp*, *An. peditaeniatus (Leicester)*, *An. subpictus Grassi*, *An. stephensi var. mysoriensis*, *An vagus* and *Mn. annulifera* (Table 4.1). Outdoor collection comprises only culicine mosquitoes in both the areas. Among indoor collections, culicines consisted of 92.8 % and 86.34 % while anophelines 7.2 % and 13.66 % in Safiabad (paddy) and Madanpur Khadar (non-paddy) area, respectively. In outdoor collection, *Cx. tritaeniorhynchus* was the predominant species (88.4 % and 88.0 % of total mosquito collection) in paddy and non-paddy area, respectively, while in indoors this species accounted for only 20.8 % and 1.09 % of total mosquito collection in paddy and non-paddy area, respectively.

The abundance (number of females/Ten Hop Cages, PTHC) of *Cx. tritaeniorhynchus* was found to be higher in paddy growing area. The PTHC densities of the species were 316.6 (September), 240 (October), 50 (November) and 6 (December), while in non-paddy area density of this species were 74, 140, 15 and 5 during September, October, November and December, respectively (Fig. 4.1).

Cx. quinquefasciatus mosquitoes rested outdoor shelters along with *Cx. tritaeniorhynchus* during cooler months (November and December). Both males and females of *Cx. tritaeniorhynchus* were found resting outdoors on the same land and aquatic vegetation. Female specimens of this species collected outdoors during September and October were in all stages of gonotrophic cycle: unfed, freshly

Fig. 4.1 Comparative abundance of *Culex tritaeniorhynchus* in outdoor collection (BPD hop cage method) in paddy and non-paddy growing areas of Northern India

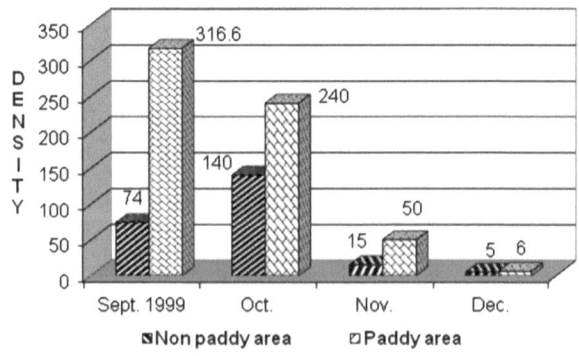

Table 4.1 Adult mosquito densities in village Madanpur khadar, Delhi and Safiabad, Haryana (September–December 1999)

Mosquito species	Madanpur Khadar Outdoor[a] Total	PTHC[c]	Indoor[b] Total	PMH[d]	Safiabad Outdoor[a] Total	PTHC[c]	Indoor[b] Total	PMH[d]
September 1999								
Cx. tritaeniorhynchus	37 (13)	74	0	0	95 (71)	316.6	2 (1)	12
Cx. fuscephala	0	0	13 (4)	78	0	0	3 (0)	18
Cx. quinquefasciatus	0	0	9 (15)	54	0	0	10 (60)	60
An. stephensi var. mys.	0	0	0	0	0	0	2 (1)	12
An. subpictus	0	0	0	0	0	0	1 (1)	6
October								
Cx. tritaeniorhynchus	112(52)	140	4(12)	20	120(3)	240	24(2)	96
Cx. pseudovishnui	0	0	0	0	0	0	2(0)	8
Cx. fuscephala	0	0	6(0)	12	0	0	0	0
Cx. quinquefasciatus	0	0	54(3)	108	0	0	27(25)	68
Cx. whitei	0	0	0	0	0	0	9(0)	36
Anopheles annularis	0	0	0	0	0	0	1(0)	4
An. peditaeniatus	0	0	0	0	0	0	2(0)	8
An. subpictus	0	0	46(28)	92	0	0	2(0)	8
An. vagus	0	0	4(0)	8	0	0	0	0
Mansonia annulifera	0	0	0	0	0	0	1(0)	4
November								
Cx. tritaeniorhynchus	15(0)	15	0	0	25 (0)	50	0	0
Cx. quinquefasciatus	7(0)	7	80(15)	80	29 (2)	56	14(8)	168
Mn. annulifera	1(0)	1	0	0	0	0	0	0
Armigeres subalbatus	0	0	0	0	1(0)	3	0	0
December								
Cx. tritaeniorhynchus	5(0)	5	0	0	3(0)	6	0	0
Cx. quinquefasciatus	15(1)	15	150(45)	600	2(0)	4	25(10)	75
Total	192		366		275		125	

[a]Hop cage collection

[b]Hand catch collection; Figures in parenthesis indicate total number of male mosquitoes of the species

[c]Density per ten hop cages

[d]Density per man hour

Table 4.2 Analysis of *Cx. tritaeniorhynchus* mosquitoes collected outdoors (Hop cage method) from study areas of Madanpur Khadar (Delhi) and Safiabad (Haryana) during September–December 1999

Months	Madanpur Khadar							Safiabad						
	Male	Female					M:F ratio	Male	Female					M:F ratio
		No.	%						No.	%				
			UF	FF	SG	G				UF	FF	SG	G	
September 1999	13	37	25	50	15	10	1:2.8	71	95	10	30	45	15	1:1.3
October	52	112	20	65	5	10	1:2.1	3	120	65	25	7	3	1:40
November	0	15	92	0	8	0	0:15	0	25	95	0	5	0	0:25
December	0	5	100	0	0	0	0:5	0	6	100	0	0	0	0:6

UF Unfed, *FF* Freshly Fed, *SG* Semigravid, *G* Gravid

fed, semigravid and gravid (Table 4.2), indicating thereby predominantly outdoor resting habit of this species in the area. *Cx. tritaeniorhynchus* mosquitoes was found resting outdoors during Japanese encephalitis outbreak in 2003 in Warangal and Karimnagar Districts, Andhra Pradesh (Das et al. 2004). Outdoor collections during November and December from both the study areas of Delhi and Haryana did not yield any male *Cx. tritaeniorhynchus*.

4.3 Identification of Various Natural Day Resting Places of *Cx. tritaeniorhynchus* in and Around Delhi

Study areas Studies were carried out in two villages of Delhi (Madanpur Khadar and Osmanpur) and one village Safiabad (Haryana). These villages are 40-km apart from each other. The village Osmanpur, the third study side is situated close to the river Yamuna in Shahdara Zone of East Delhi (Map 4.1) and is partially surrounded by vast marshy area with luxuriant growth of water hyacinth nearly round the year. Land vegetation in the area comprises wild grasses only.

To locate the preferential outdoor resting habitat of vector species, adult mosquito collection was carried out employing BPD hop cage method from land and aquatic vegetation in the study areas of Delhi and Haryana. Land vegetation in the study villages were mainly *Sorghum bicolor* (jowar-millets), *Trifolium alexandrinum* (berseem), 30 cm high wild shrubs and seasonal vegetable: mustard, carrot, brinjal, marigold (flower) and aquatic vegetation included hyacinth plant (*Eichhornia crassipes*) and elephant grass.

Usually, less than 2 min/per hop cage were required to collect mosquitoes from their natural day resting shelters among aquatic and land vegetation following the present method. Table 4.3 gives the preferential daytime natural outdoor resting

Table 4.3 Preferential day-time outdoor resting natural shelters of adult mosquitoes in and around Delhi during Jan-December 2000 (Hop cage method)

Habitat			1	2	3	4	5	6	7	8	9	10	No.
			(*)	(*)	(*)	(*)			(*)		(*)	(*)	(**)
Aquatic	Water hyacinth	1	+			+					+		3
veg.	Elephant grass	2	+			+							2
Land	GV in Jowar field	3	+	+					+	+		+	5
veg.	GV in Bajra field	4	+	+				+	+	+	+		6
	Carrot	5	+										1
	Merigold	6	+		+	+	+						4
	Brinjal	7	+										1
	Cauliflower	8	+										1
	Cabbage	9											
	Spinach	10	+		+	+	+					+	5
	Paddy stalk	11											
	Wheat stalk	12											
	Hemp (wild veg.)	13	+		+								2
Preference of resting sites for the species		1,3,4,2, 10, 6,7, 5,13,8	3,4	6, 10, 13	6,1, 10,2		6,10	3,4	3,4	1,4	3,10		

1, *Culex tritaeniorhynchus,* 2, *Cx. pseudovishnui,* 3, *Cx. fuscocephala,* 4, *Cx. quinquefasciatus,* 5, *Cx. whitei,* 6. *Anopheles annularis,* 7. *An. peditaeniatus,* 8, *An. pulcherrimus,* 9, *Mansonia annulifera* and 10, *Armigeres subalbatus*;

GV = Ground vegetation; + = Denotes the presence of species in particular habitat

* = Vectors of JEV in India; ** = No. of species in each habitat

shelters of adult mosquitoes in the study area. A total of 1,605 female and 596 male mosquitoes comprising ten species in four genera, viz. *Cx. tritaeniorhynchus*, *Cx. pseudovishnui*, *Cx. quinquefasciatus*, *Cx. whitei*, *Cx. fuscocephala*, *Anopheles annularis*, *An. peditaeniatus*, *An. pulcherrimus*, *Mansonia annulifera* and *Armigeres subalbatus* were collected from their natural day resting shelters among vegetation during the study period. Of these, the first species was found nearly in all the vegetation searched. The most prevalent mosquito species resting during the day in outdoor shelters was *Cx. tritaeniorhynchus* (89 %). These were mostly hop cage collections (97.41 %). Concurrent hand catch collections failed to yield many mosquitoes. The most preferred natural day resting shelters for *Cx. tritaeniorhynchus* was the dense bushy vegetation of water hyacinth (Chap. 2: Fig. 2.3b) just above the water level (in Osmanpur and Madanpur Khadar village) and ground vegetation up to a height of 30–45 cm in millet field (in Safiabad village) followed by elephant grass, marigold, spinach, carrot, etc. Least-preferred day resting shelters include hemp field, brinjal and cauliflower. Collections made from majority of wild vegetation, cabbage, paddy and wheat stalk did not yield any specimen of adult specimen, similar observation was noted elsewhere (Kulkarni and Rajput 1988).

4.4 Comparative Efficacy of Drop Net, Hand Catch and BPD Hop Cage Method for Monitoring JE Vector Abundance

The field efficacy of the BPD hop cage method was tested in six villages and one periurban localities of Delhi and Haryana state of India. These villages were selected as they had the potential day resting sites of mosquitoes on land and aquatic vegetation. Comparative efficacy of hop cage and hand catch method for sampling outdoor resting population of mosquitoes was carried out at monthly interval in three villages: two villages of Delhi (Osmanpur, Madanpur Khadar) and one village Safiabad (Sonipat district: Haryana). Whereas comparative efficacy of the three methods, viz: hop cage, drop net and hand catch method was undertaken in five study villages of district Karnal (JE endemic district of Haryana state of India).

Area covered & Mosquito density measurement by hop cage, drop net and hand catch method: For comparative efficacy, mosquitoes were collected simultaneously from the same vegetation by hop cage, drop net and hand catch method, taking utmost care that same area was not searched by two methods and area covered in searching for mosquitoes was 50 ft^2.

The hop cage was moved over low level ground vegetation through a distance of 50 ft^2 and the density was measured as average number of mosquitoes collected in ten hop cages (PTHC). Each hop on vegetation covers an area of 1 ft^2 and in one attempt the hop cage was moved through a distance of 5′. The area covered by 10 such attempts equals $[(1′ \times 1′) \times 5 \text{ ft} = 5 \text{ ft}^2] \times 10 = 50 \text{ ft}^2$. Although, WHO recommended drop net size is $6′ \times 6′ \times 6′$, in order to conform with the uniformity in terms of similar area to be screened by all the three candidate sampling methods, a slightly smaller drop net $5′ \times 5′ \times 5′$ was used twice in each type of vegetation (Fig. 4.2) and the mosquito density was measured as average number of mosquitoes collected in two drop nets. Therefore, area covered by two such attempts equals $[(5′ \times 5′) = 25 \text{ ft}^2] \times 2 = 50 \text{ ft}^2$.

Outdoor resting mosquitoes were also collected from land vegetation by aspirator tube method (hand catch) covering an area of 50 ft^2 and the density of mosquito species expressed as average number collected per man hour (PMH). Collection made in and Delhi was immediately transported to Vector Ecology Laboratory, national institute of communicable diseases (NICD) and those from

Fig. 4.2 Collection of outdoor resting population of *Cx. tritaeniorhynchus* by **a** Drop net from dense vegetation of Jowar (90 cm long), **b** Hop cage method (hop cage is not visible in this case)

study areas of District Karnal were transported to field laboratory at Karnal town. Collected mosquitoes were identified and processed.

Comparative efficacy of BPD hop cage and hand catch method undertaken in aquatic vegetation in Osmanpur (Delhi) revealed that *Cx. tritaeniorhynchus* was collected throughout the year using hop cage method with density ranging from 1–120 PTHC (Fig. 4.3a). During the same time: in another portion of the same habitat, the per man hour density (PMHD) for the species ranged from 1 to 4 using hand catch method (Fig. 4.3b). Similarly, among land vegetation in Safiabad village, density of *Cx. tritaeniorhynchus* ranged from 1–280 PTHC (Fig. 4.3c) as against 0–8 PMH by the conventional hand catch method (Fig. 4.3d). During month of August (peak monsoon season), density of the species using hop cage and hand catch method in Safiabad village was 280 PTHC and 8 PMH, respectively. Study at Madanpur Khadar (Delhi) situated on the bank of Agra canal revealed: PTHC density of *Cx. tritaeniorhynchus* ranged from 2 to 84 as against 0–10 PMH by the conventional hand catch method (Table 4.4; Fig. 4.3e, f).

Adult mosquito collection using BPD hop cage method during monsoon and post-monsoon months revealed that a favourable mosquitogenic conditions prevails in and around Delhi. It is also noteworthy that population fluctuation of *Cx. tritaeniorhynchus* can be monitored round the year in outdoor vegetation employing hop cage method. In contrast, conventional hand catch method failed to reveal such fluctuation in the population abundance of the species. *Cx. tritaeniorhynchus* is known to be a zoophilic species and prefers to feed on bovine and swine population. However, under high density situation (during monsoon and post-monsoon months), the species turns into an indiscriminant feeder and it is

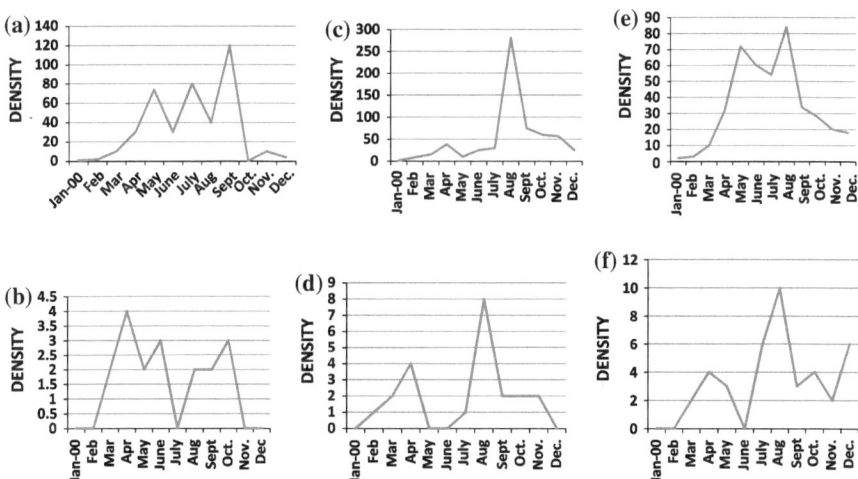

Fig. 4.3 Comparative efficacy of hand catch and BPD hop cage method in sampling outdoor resting population of *Culex tritaeniorhynchus* in study villages during 2000. Hop cage method: **a** Osmanpur village; **c** Safiabad; **e** Madanpur Khadar. Hand catch method: **b** Osmanpur; **d** Safiabad; **f** Madanpur Khadar

Table 4.4 Comparative efficacy of BPD hop cage and Hand catch method in monitoring density of *Cx. tritaeniorhynchus* (females) under outdoor situation in and around Delhi during (January–December 2000)

MONTHS	SAFIABAD (HARYANA)			OSMANPUR (DELHI)			MADANPUR KHADAR (DELHI)		
	% of total mosquito collected	Density Hop cage (PTHC)	Hand catch (PMH)	% of total mosquito collected	Density Hop cage (PTHC)	Hand catch (PMH)	% of total mosquito collected	Density Hop cage (PTHC)	Hand catch (PMH)
Jan. 2000	100	1	0	66.6	1	0	50	2	0
Feb	100	9	1	5.2	2	0	25	3	0
Mar	50	15	2	20	10	2	18	10	2
Apr	50	38	4	75	30	4	50	32	4
May	100	10	0	100	74	2	100	72	3
June	100	25	0	100	30	3	75	60	0
July	100	30	1	100	80	0	100	54	6
Aug	99	280	8	100	40	2	100	84	10
Sept	100	75	2	100	120	2	95	34	3
Oct.	100	60	2	100	10*	3	100	28	4
Nov.	95	57	2	33.3	10	0	80	20	2
Dec.	94	25	0	51.8	4	0	75	18	6

PTHCD = Per Ten Hop Cage Density PMHD = Per Man Hour Density

* = Collected just after rain

then a fraction of the population may be diverted to human host increasing man–mosquito contact. It is, therefore, mandatory for a sampling tool used for a predominantly zoophilic species to be adequately sensitive to detect sharp increase in population density of such species in order to alert the local authorities to undertake appropriate intervention methods to prevent the disease outbreak.

Comparative sensitivity of hop cage, drop net and hand catch method in measuring the density of *Cx. tritaenorhynchus* was undertaken in land vegetation like jowar during pre-monsoon and monsoon months, berseem during post-monsoon month in study villages of Karnal District. Density of *Cx. tritaenorhynchus* was 57.8 per ten hop cages, 27 per two drop nets and 0.0 per man hour during pre-monsoon; 91.5 and 79.5 per ten hop cages, 26.3 each per two drop nets and 1 and 8 per man hour during monsoon months of July and September and 104 per ten hop cages, 51 per two drop nets and 6 per man hour during post-monsoon month of November using hop cage, drop net and hand catch method, respectively (Fig. 4.4).

In ten hopping on vegetation using hop cage method and retrieving the collected mosquitoes required 15–25 min depending on abundance of mosquito species and the efficiency of the collector. In contrast, using two drop nets three collectors require 45–60 min for sampling mosquitoes only from land vegetation and that too when the vegetation is not more than 3 ft long. Hop cage method is a more sensitive tool than conventional drop net and hand catch method for sampling outdoor

Fig. 4.4 Comparative efficacy of hop cage, drop net and hand catch method for sampling outdoor resting population of *Cx. tritaeniorhynchus* from land vegetation

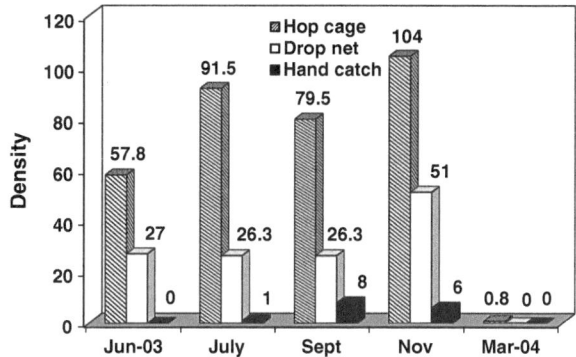

resting population of *Cx. tritaenorhynchus* from both land and aquatic vegetation. In contrast, in conventional drop net method, the nets are very large, inconvenient to operate in the field, very difficult to transport over long distance, less coverage, requires more man power and economy to incorporate the tool in JE vector surveillance programme. Conventional hand catch method also was found to be unsuitable for sampling outdoor resting population of JE vectors due to presence of innumerable, minute and inaccessible natural resting sites provided by both aquatic and land vegetation in the study villages. Hop cage cloth gets dirty particularly when used in aquatic environment due to algae, muddy soil, etc., and should be removed periodically from the cage frame and washed or replaced.

Changes in densities of a mosquito species resting in different types of vegetation at different times of the year can be obtained if the collecting technique is standardised. Density of mosquito species resting in vegetation using sweep net is calculated by average number of mosquitoes caught per net or by per ten nets (Service 1976). Sweep netting is an efficient sampling technique for non-medical insects resting on different types of vegetation.

4.5   Use of BPD Hop Cage Method in JE Vector Surveillance and Outbreak Investigation of JE/AES

The data presented in this section allow (i) suggesting the utility of BPD hop cage method in JE vector surveillance, and (ii) its use in outbreak investigation of AES/ Japanese encephalitis.

4.5.1 JE Vector Surveillance

Hop cage method was used in many aspects of JE vector bionomics studies undertaken in both JE endemic and non-endemic areas of the country, viz: two

Fig. 4.5 Use of BPD hop cage method in monitoring JE vector abundance in JE vector surveillance and in outbreak investigation of JE/AES carried out in different states of India

non-endemic areas: Delhi and Sonipat district, Haryana; two endemic areas: Karnal district, Haryana and Saharanpur district, Uttar Pradesh (Fig. 4.5).

4.5.1.1 Identification of Vector Species

Mosquitoes caught in hop cage are alive and do not tend to become denuded of scales, hence are ideal for taxonomic studies.

4.5.1.2 Monitoring JE Vector Abundance

It was interesting to note that in areas where *Cx. tritaeniorhynchus* (principal vector of JE in India) primarily rests outdoors and BPD hop cage method is the most suitable tool for monitoring its abundance round the year (Chap. 5: Fig. 5.1, 5.2; Chap. 6: Fig. 6.5; Chap. 7: Fig. 7.10).

Fig. 4.6 Population structure of *Cx. tritaeniorhynchus* in outdoor collection (hop cage)

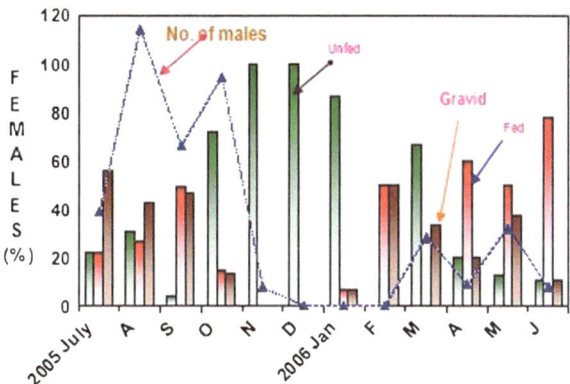

4.5.1.3 Determination of True Exophilic Nature of JE Vector Species in an Area

Cx. tritaenorhynchus mosquitoes collected from land vegetation by employing hop cage method revealed that with the exception of winter months (November–February), female mosquitoes collected outdoors were in all stages of gonotrophic cycle: unfed, freshly fed, semigravid and gravid (Fig. 4.6). Use of hop cage method revealed true outdoor resting habit of *Cx. tritaenorhynchus* with entire population sharing the same resting habitat among land/aquatic vegetation in Delhi, Haryana, Uttar Pradesh, Uttaranchal and Andhra Pradesh state of India.

4.5.1.4 First Time Detection of Overwintering Phenomenon in *Cx. tritaeniorhynchus* Mosquitoes in India

During November, wild caught *Cx. tritaeniorhynchus* females in outdoor collection (hop cage method) were unfed (Fig. 4.7). As night temperature dropped in the area, females stopped taking blood meal and overwinter resting in available local vegetation from mid-November to mid-February. Thereafter, due to gradual rise in night temperature, overwintering females came out of their overwintering shelter, started taking blood feed and population builds up slowly. In an earlier study (Das 2003), overwintering phenomenon in *Cx. tritaenorhynchus* females was reported first time from Delhi. This was the first report of overwintering phenomenon in *Cx. tritaenorhynchus* mosquitoes from India.

4.5.1.5 Feeding Preference of Mosquitoes

Female mosquitoes collected employing hop cage method is unbiased as they are the resting population. Vector species (*Cx. tritaeniorhynchus*) is known to have a very long (in unfavourable condition >5 km.) flight range and upon subjected to mosquito blood meal analysis reveals the actual host preference for the species

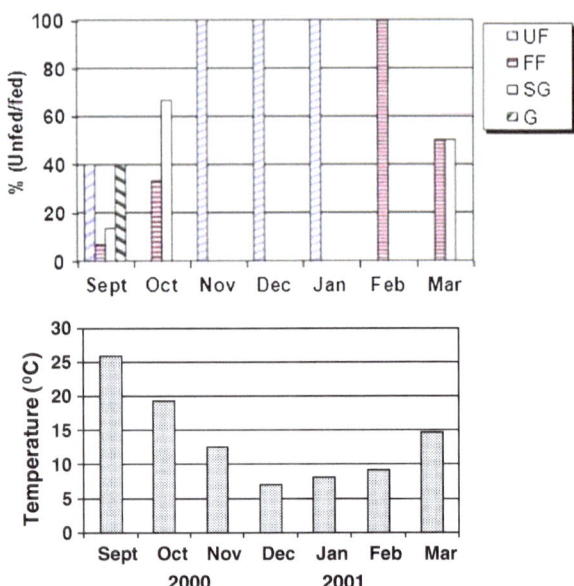

Fig. 4.7 *Cx. tritaenorhynchus* population overwinters in outdoor vegetation (as unfed females) during winter months in Northern India

including avian source. In contrast, dusk collection though may yield adequate vector species, mosquito blood meal analysis of such population in all probability indicate preference of the host species in the vicinity of the collection site.

4.5.1.6 JE Virus Infection in Vector Species for Detection of Early Warning Signal

JE virus infection was detected (ELISA method) in *Cx. tritaeniorhynchus* (collected using hop cage method) 2 months before the appearance of human cases in Saharanpur district (UP). This was an early warning signal for initiating appropriate vector control/management measures to prevent JE outbreak.

4.5.2 Outbreak Investigation of AES/JE

During entomological investigation of an outbreak of Acute Encephalitis syndrome (AES)/JE in JE endemic area, main thrust should be to collect adequate sample of adult JE vector species in indoor (conventional hand catch/total catch method) and outdoor situation (BPD hop catch method) for ascertaining the main JE vector density pattern in the affected area, their preferred resting places where appropriate vector control measures can be applied to interrupt the transmission.

Hop cage method was used in JE/AES outbreak investigation studies in many states of the country (Fig. 4.5). These are JE endemic districts: Karnal and Kurukshetra district (Haryana); Gorakhpur, Kushinagar and Saharanpur districts

Table 4.5 Adult mosquito densities in two JE affected villages of Gorakhpur district, Uttar Pradesh (September 2004)

Mosquito species	Village: Mirjapur (PHC: Khorabar)								Village: Bargatrahin (Bhat hat)	
	Outdoor				Indoor				Evening	
	Land vegetation		Aquatic vegetation		Human dwelling		Cattle shed		Human dwelling	
	Total*	PTHC♣	Total	PTHC	Total	PMH♦	Total	PMH	Total	PMH
Culex tritaenio-rhyn-chus	12 (100)	60.0	1 (100)	5.0	-	-	-	-	20 (36.4)	20.0
Cx. bitaeniorhyn-chus	-	-	1 (100)	5.0	-	-	-	-	1 (1.8)	1.0
Cx. gelidus	-	-	-	-	-	-	-	-	3 (5.6)	3.0
Cx. quinquefascia-tus	-	-	-	-	18 (100)	72.0	6 (66.7)	12.0	1 (1.8)	1.0
Cx. perplexux	-	-	-	-	-	-	-	-	1 (1.8)	1.0
Anopheles annularis	-	-	-	-	-	-	2 (22.2)	4.0	2 (3.6)	2.0
An. peditaeniatus	-	-	-	-	-	-	-	-	8 (14.5)	8.0
An. subpictus	-	-	-	-	-	-	-	-	1 (1.8)	1.0
An. vagus	-	-	-	-	-	-	-	-	1 (1.8)	1.0
Mansonia indiana	-	-	-	-	-	-	-	-	17 (30.9)	17.0
Armigeres subalbatus	-	-	-	-	-	-	1 (11.1)	2.0	-	-
Total	12		2		18		9		55	

*= Female; ♣ = Per Ten Hop Cages ♦ = Per Man Hour Figures in parenthesis indicate prevalence of a mosquito species in this collection

(UP state of India); Nainital and Udham Singh Nagar (Uttaranchal state); Warangal and Karimnagar districts (Andhra Pradesh state). During (2003) viral encephalitis outbreak in Warangal and Karimnagar Districts of Andhra Pradesh, adult mosquito collection employing BPD hop cage method in low-level wild shrubs and seasonal cash crops revealed very high density of *Cx. tritaeniorhynchus* as compared to its poor density in indoor situations in cattle sheds and human dwellings. The density of *Cx. tritaeniorhynchus* was found to be high on land vegetation in all the three study villages: Medipathy (50.0 PTHC), Pegadapalli (87.5 PTHC) and Uppal (71.6 PTHC). The indoor density of the species in cattle sheds was 2.3 PMH in village Uppal (Table 4.5). Similar trend was detected in outbreak investigation of Japanese encephalitis in other JE endemic areas of Northern India (Fig. 4.8).

Fig. 4.8 Abundance of
Culex tritaeniorhynchus
in outdoor collection (hop
cage method) and indoor
collection (hand catch
method) during outbreak
investigation of JE in Andhra
Pradesh, Uttar Pradesh and
Haryana

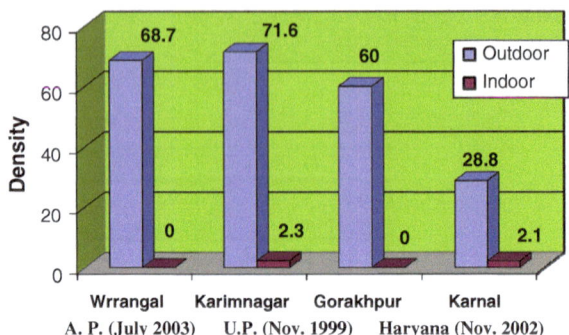

BPD hop cage method was included in Chap. 5 "Entomological surveillance" in Government of India Guidelines for surveillance of AES with special reference to Japanese encephalitis (Government of India guidelines 2006).

Among mosquitoes that rest outdoors, vectors of JE are most important as they mainly rest outdoors in vegetation in India and transmit the disease in over 155 districts in the country (Source: National Vector Borne Disease Control Programme). Among the exophilic culicine mosquitoes, three species, *Cx. tritaeniorhynchus*, *Cx pseudovishnui* and *Cx. vishnui* are known to be involved in the transmission of JE virus in India (Reuben 1971; Reuben and Gajanana 1997). However, *Cx. tritaeniorhynchus* has been mainly involved in the outbreaks of JE in central, northern and southern India (Reuben 1971; Carey et al. 1968; Mathur et al. 1981; Kaliwal 1991; Pant et al. 1994; Gajanana et al. 1997; Philip Samuel et al. 2000). The species though exophilic in its resting behaviour (Das et al.2004; Kulkarni and Rajput 1988; Khan et al. 1997), has been reported in low densities in indoor situation too during JE outbreak (Sharma et al. 1991; Dash et al. 2001).

Hop cage method is simple as well as convenient method to study the abundance of *Cx. tritaenorhynchus*, and can be a useful method for vector surveillance in JE endemic as well as non-endemic areas of the country with similar pattern of vector ecology. Research is also needed to know the index vegetation in diverge ecological situation in the country where hop cage method can be used to provide the most sensitive indication of population abundance. BPD hop cage used in the entire period of 9-year study and the methodology adopted provide ample opportunity: (i) to monitor JE vector density in areas with known high prevalence of the disease, in order to detect sharp rise in population density of JE vector species, (ii) to know seasonal prevalence of proven/suspected JE vectors in an area, (iii) to collect mosquitoes for virus isolation, (iv) to determine range of host preference of JE vectors including avian source, etc. The hop cage is an ordinary mosquito cage and the methodology adopted is also very simple and does not involve any extra budget for the programme. With a little training, the entomological staff available with the programme and the research institutes would be in a position to detect sharp increase in population density of *Cx. tritaeniorhynchus* in outdoor situation in their area. This is an essential component in developing an early warning signal for the disease, so that timely preventive measures may be taken by the local authorities to prevent outbreak of the disease.

References

Arunachalam N, Philip Samuel P, Hiriyan J, Thenmozhi V et al (2002) Vertical transmission of Japanese virus in *Mansonia* species, in an epidemic-prone area of southern India. Ann Trop Med Parasitol 96:419–420

Carey DE, Reuben R, Myres RM, Pavri KM et al (1968) Japanese encephalitis studies in Vellore, South India. I: virus isolation from mosquitoes. Indian J Med Res 56:1309–1318

Das BP (2003) *Chilodonella uncinata*—a protozoa pathogenic to mosquito larvae. Curr Sci 85:483–489

Das BP (2009) BPD hop cage method—a new device of collecting mosquitoes for effective JE vector surveillance. Invent Intell 44:24–25

Das BP, Lal S, Saxena VK (2004) Outdoor resting preference of *Culex tritaeniorhynchus,* vector of Japanese encephalitis in Warangal and Karim Nagar district, Andhra Padesh. J Vector Borne Dis 41:32–36

Das BP, Rajagopal R, Akiyama J (1990) Pictorial key to the species of Indian Anopheline mosquitoes. J Pure Appl Zool 2:131–162

Das BP, Sharma SN, Kabilan L, Lal S et al (2005) First time detection of Japanese Giles, 1901, from Karnal district of Haryana state of India. J Commun Dis 37:131–133

Dash A, Chhotray GP, Mahapatra N, Hazra RK (2001) Retrospective analysis of epidemiological investigation of Japanese encephalitis outbreak occurred in Rourkela, Orissa, India. Southeast Asian J Trop Med Public Health 32:137–142

Gajanana A, Rajendran R, Philip Samuel P, Thenmozhi V et al (1997) Japanese encephalitis in south Arcot district, Tamil Nadu, India: a three year longitudinal study of vector abundance and vector infection frequency. J Med Entomol 34:651–659

Government of India (2006) Dte. of National Vector Borne Disease Control Programme. Guidelines for surveillance of Acute Encephalitis Syndrome (with special reference to Japanese encephalitis). D.G.H.S., Ministry of Health & Family Welfare

Gupta N, Hossain S, Lal R, Das BP et al (2005) Epidemiological profile of Japanese encephalitis outbreak in Gorakhpur, U.P. in 2004. J Commun Dis 37:145–149

Kaliwal MB ((1991) Entomological investigation of Japanese encephalitis outbreak during 1991 in Goa. Bull Vector Borne Dis 3:13–17

Khan SA, Dutta P, Narain K et al (1997) Studies on day-time resting habits of JE vector mosquitoes in upper Assam with a note on insecticide susceptibility status. J Commun Dis 29:367–370

Kulkarni SM, Rajput KB (1988) Daytime resting habitats of Culicine mosquitoes and their preponderance in Bastar District, Madhya Pradesh, India. J Commun Dis 20:280–286

Mathur KK, Bagchi SK, Sehgal CL, Bhardwaj M (1981) Investigation on an outbreak of Japanese encephalitis in Raipur, Madhya Pradesh. J Commun Dis 13:257–265

Murthy S, Singh TG, Arunachalam N, Philip Samual P (2000) Epidemiology of Japanese encephalitis in Andhra Pradesh, India—a brief overview. Trop Biomed 17(2):97–100

Pant U, Ilkal MA, Somen RS, Shetty PS et al (1994) First isolation of Japanese encephalitis virus from the mosquito *Culex tritaeniorhynchus* Giles, 1901 (Diptera: Culicidae) in Gorakhpur District, Uttar Pradesh. Indian J Med Res 99:149–151

Philip Samuel P, Hiriyan J, Gajanana A (2000) Japanese encephalitis virus infection in mosquitoes and its epidemiological implications. ICMR Bull 30:37–43

Rajagopalan PK, Menon PKB, Panicker KN (1978) An ecological appraisal of the mosquitogenic conditions and probable causes of the 1978 epidemic of 'Encephalitis' in Tirunelveli district, Tamil Nadu. J Commun Dis 10:157–164

Reuben R (1971) Studies on the mosquitoes of North Arcot District, Madras state, India. Part 1. Seasonal densities. J Med Entomol 8:119–126

Reuben R, Gajanana A (1997) Japanese encephalitis in India. Indian J Paediatr 64:2433–2451

Reuben R, Tewari SC, Hiriyan J, Akiyama J (1994) Illustrated key to genera of *Culex (Culex)* associated with Japanese encephalitis in Southeast Asia (Diptera: Culicidae). Mosq Syst 26:75–96

Saxena VK, Baig MH, Bhardwaj M, Rajagopal R (1986) Entomological investigations of Japanese encephalitis outbreak in Gorakhpur and Deoria districts of Uttar Pradesh. J Commun Dis 18:219–221

Service MW (1976) Mosquito ecology: field sampling methods. Applied Science Publishers Ltd., London, pp 583

Sharma RC, Saxena VK, Bharadwaj M et al (1991) An outbreak of Japanese encephalitis in Haryana–1990. J Commun Dis 23:168–169

Sirivanakarn S (1976) Medical Entomology studies III. A revision of the subgenus *Culex* in the Oriental Region (Diptera: Culicidae). Contrib Am Entomol Inst (Ann Arbor) 12(2):1–272

World Health Organization (1975) Methods and techniques. In: Manual on practical entomology in malaria, Part II. Geneva, pp 186

Chapter 5
Ecology of *Culex tritaeniorhynchus* in and Adjoining Areas of Delhi, Non endemic Area in Northern India, with Special Reference to *Chilodonella uncinata* as a Bio-control Agent

Abstract *Culex tritaeniorhynchus* breeds extensively in paddy fields in villages located in Sonipat district, Haryana close to Nerela border, North Delhi from where field studies carried out at NICD led to the discovery of a tiny microbe *Chilodonella uncinata* (protozoan parasite) that kills JE vector larvae growing in paddy fields and burrow pits. In contrast, in the absence of paddy fields in East and South Delhi vast marshes and disused ground pools with profuse growth of water hyacinth were the most preferred breeding habitats for *Cx. tritaeniorhynchus* mosquitoes. Larvae of this species in Delhi were found to be tolerant to *Ch. uncinata* infection. Though Delhi and adjoining villages of Haryana are not endemic to JE, sporadic cases were reported from many areas of Delhi during 2002–2005 and recently during 2011. Aim of this chapter was to use the BPD hop cage method in the ecological study of *Cx. tritaeniorhynchus,* principal vector of JE in India in and adjoining areas of Delhi, non endemic area in Northern India. Differential role of *Chilodonella uncinata* was demonstrated on the population abundance of *Cx. tritaeniorhynchus* in diverse ecological situation in and around Delhi. Situation specific JE vector control/management strategies are suggested to prevent transmission of the disease.

5.1 Introduction

Delhi is situated between 28°39′N latitude and 77°13′E longitude and shares its borders with Haryana and Uttar Pradesh. The approximate count of population residing in Delhi is 1,38,50,507 according to the 2001 Census. The main crops grown in and around Delhi are wheat, jowar, bajra, and paddy. Farmers in Delhi, Haryana, and Uttar Pradesh are found to produce a large amount of green fodder crops such as berseem for cattle feed. These crops demand relatively little attention, allowing farmers to focus their efforts on cultivating other crops produce. The three main sources of irrigation in Delhi are canals, wells, and tube wells. Totally, 92–95 % of the net irrigated area were by wells/tube wells (as per Economic survey of Delhi 2001–2002).

Delhi has a hot and humid climate with long and extremely hot summer season, from early April to mid October with monsoon season in between. The average

B. P. Das, *Mosquito Vectors of Japanese Encephalitis Virus from Northern India*, SpringerBriefs in Animal Sciences, DOI: 10.1007/978-81-322-0861-7_5, © The Author(s) 2013

Table 5.1 State-wise cases and deaths in India due to suspected Japanese encephalitis (2000–2005)

Affected States/ UTs	2000		2001		2002		2003		2004		2005	
	C	D	C	D	C	D	C	D	C	D	C	D
Andhra Pradesh	343	72	33	4	22	3	329	183	7	3	34	0
Assam	158	69	343	200	472	150	109	49	235	64	145	52
Bihar	77	19	48	11	8	1	6	2	85	28	192	64
Chandigarh	0	0	0	0	4	0	0	0	0	0	0	0
Delhi	0	0	0	0	1	0	12	5	17	0	6	0
Goa	15	3	6	2	11	0	0	0	0	0	4	0
Haryana	74	43	47	22	59	40	104	67	37	27	46	39
Karnataka	438	45	206	14	152	15	226	10	181	6	122	10
Kerala	164	2	128	5	0	0	17	2	9	1	1	0
Maharashtra	82	0	126	1	119	16	475	1s15	22	0	51	0
Manipur	0	0	0	0	2	1	1	0	0	0	1	0
Punjab	0	0	0	0	10	2	0	0	0	0	1	0
Tamil Nadu	4	0	0	0	0	0	163	36	88	9	51	11
Uttar Pradesh	1,170	252	1,005	199	604	133	1,124	237	1,030	228	6061	1500
West Bengal	148	50	119	21	301	105	2	1	3	1	12	6
Grand Total	2,593	556	2,061	479	1,765	466	2,568	705	1,695	367	6,727	1,682

Source Data obtained from the National Vector Borne Disease Control Programe

temperature during summer ranges from 25 to 46 °C. The area is fed by southwest monsoon. The average annual rainfall is approximately 718 mm (28.1 inches), most of which falls during monsoon in July, August, and September. Winter Season starts from the end of November and continues till February to March. The cold waves from the Himalayan region make the winters in Delhi very chilly. Temperatures fall substantially down to as low as 3–4 °C at the peak of winter in January.

Outbreaks of JE have occurred regularly since 1990 in four districts, viz. Karnal, Kaithal, Panipat, and Kuruskhetra of Haryana, but none had occurred up to 2001 from adjacent Delhi region (Table 5.1) which has reported sporadic cases of JE during 4 years (2002–2005). Villages located in the border district of Haryana close to Narela zone of north Delhi showed enormous breeding potential for JE vectors and *Chilodonella uncinata,* protozoan parasite was found to cause high (85–100 %) mortality JE vector larvae growing in paddy fields, ponds, and burrow pits (Das 2003). During 2000–2001, many slum areas of south Delhi were evacuated and resettled at new sites of Madanpur Khaddar in between Agra canal and River Yamuna where JE vectors were found in abundance (Chap. 4). In view of the above, a 2-year ecological study was carried out between 2000 and 2001 to determine seasonal prevalence, resting, and feeding habit including breeding characteristics of *Cx. tritaeniorhynchus* mosquitoes in and adjoining areas of Delhi.

Monthly visits were made in three selected villages (Chap. 4: Map 4.1), viz. Osmanpur and Madanpur Khadar (Delhi) and Safiabad village (Sonipat district: Haryana). Adult mosquitoes were collected during 09:00–13:00 h from outdoor resting habitats in land vegetation (jowar, mustard, and berseem) crop and aquatic vegetation (water hyacinth) using BPD hop cage method (Das 2000, 2009). Adult mosquitoes were also collected during 08:00–10:00 h from domestic animal shelters and inside houses including living room with the help of sucking tube and torch light. Animal shelters were mostly closed type made up of bricks with cement slab roof. Collections were immediately transported to Vector Ecology Laboratory, national institute of communicable diseases (NICD), sorted according to species, identified based on standard keys (Sirivanakarn 1976; Reuben et al. 1994; "Pictorial key to common species of *Culex (Culex)* mosquitoes associated with Japanese encephalitis virus in India"—Chap. 3 of this book; Das et al. 1990) and female abdominal condition (unfed, freshly fed, semi gravid, and gravid) was recorded.

5.2 Seasonal Abundance of *Culex tritaeniorhynchus* in and Around Delhi

Adult mosquito survey revealed *Culex tritaeniorhynchus* was the most abundant JE vector species in and around Delhi. Figure 5.1 shows monthly averages for *Cx. tritaeniorhynchus* taken in outdoor and indoor collections at Madanpur Khadar, Osmanpur and Safiabad, during January 2000–December 2001. Other JE vector species viz. *Cx. vishnui* and *Cx. pseudovishnui* were taken in small numbers during monsoon season only. It will be seen that abundance of the species could be monitored round the year under outdoor situation among land and aquatic vegetation at Madanpur Khadar while in indoor situation they could be collected in large numbers only during monsoon months of July, August, and September (Fig. 5.1a) with peak density observed in August (224 PMH). More or less similar type of behavior in seasonal prevalence was observed at Osmanpur (Fig. 5.1b) and Safiabad (Fig. 5.1c) where females of *Cx. tritaeniorhynchus* were collected round the year in outdoor resting places. The species was collected from indoor resting places during monsoon months (July–September) at Osmanpur and from July to October at Safiabad (Fig. 5.1c). A similar trend was observed in Vellore District, Tamil Nadu, Southern India where outdoor resting abundance of *Cx tritaeniorhynchus* was much higher that observed indoor resting situations (Reuben 1971). Catches for the species among outdoor resting places in Jowar field (land vegetation) from paddy growing area at Safiabad were significantly higher (280 PTHC) during peak abundance period (August) than that (120 PTHC) from water hyacinth (aquatic vegetation) from nonpaddy area at Osmanpur.

In earlier studies from rural areas of Delhi, females of *Cx. tritaeniorhynchus* were collected only from indoor resting places during monsoon and postmonsoon season (Menon and Rajagopalan 1976; Rahman et al. 1978). It is evident from the present study that natural outdoor resting places of the species were not detected from this part of the country in earlier studies.

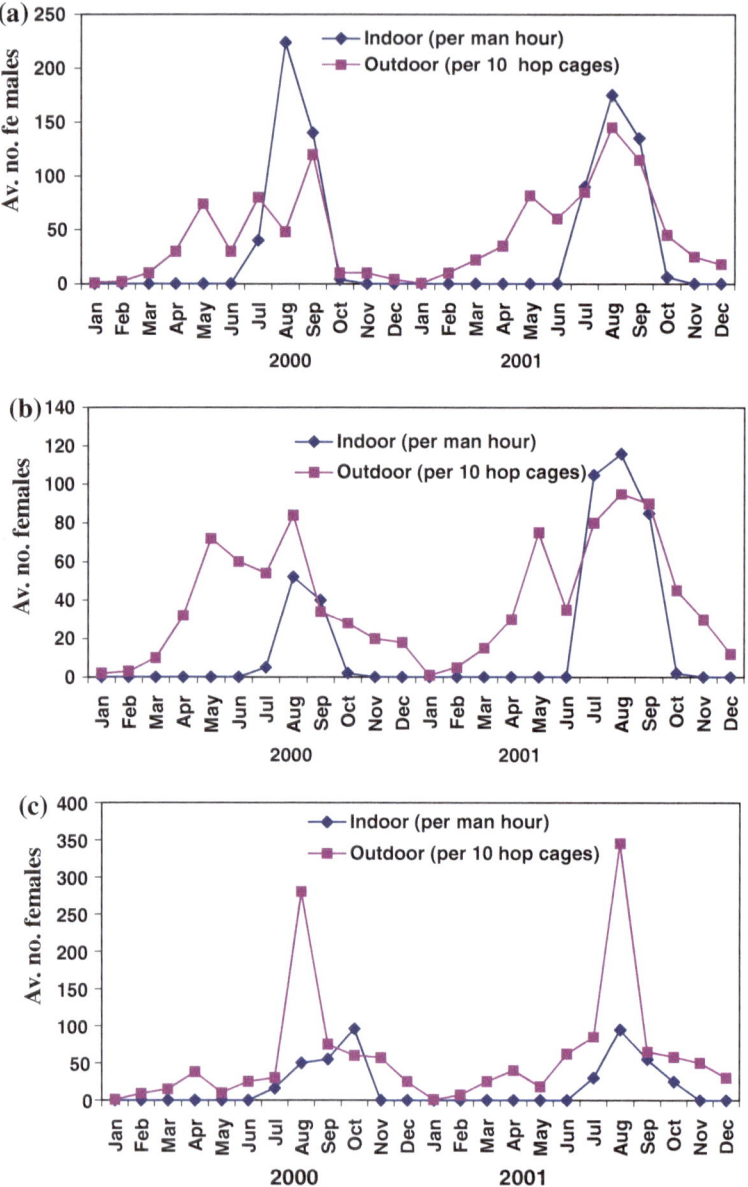

Fig. 5.1 Seasonal abundance of *Cx. tritaeniorhynchus* in outdoor collection (BPD hop cage method) and indoor collection (Hand catch method). **a** Madanpur Khadar (Delhi). **b** Osmanpur (Delhi). **c** Safiabad (Haryana)

Population curve of *Cx. tritaenorhynchus* in outdoor collection showed bimodal pattern of peak occurrence. At Osmanpur, a small summer peak was observed in May (82 PTHC) and a higher monsoon peak (145 PTHC) in August (Fig. 5.2a). *E. crassipers* (water hyacinth) promotes profuse breeding of *Cx. tritaenorhynchus*.

Fig. 5.2 Population curve of *Culex tritaeniorhynchus* in outdoor collection (Nonendemic area). **a** Aquatic vegetation (Delhi). **b** Land vegetation (Sonepat district, Haryana)

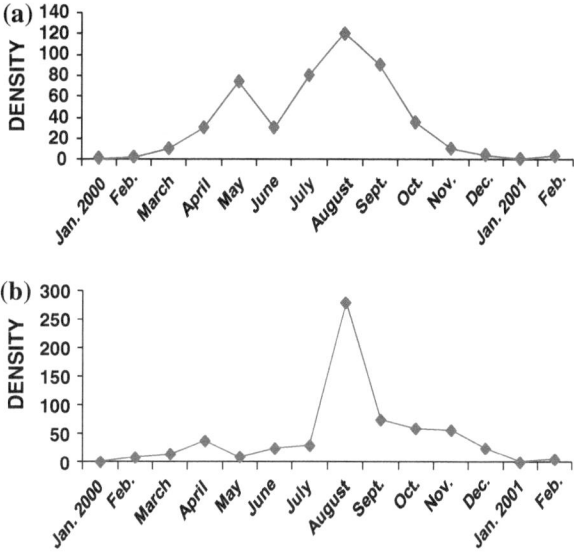

Moreover, during dry and hot summer month, numerous minute sites among the fibrous roots of the plant provide moist cool enormous microclimate as resting sites for the species (Chap. 2: Fig. 2.3b). In village Safiabad, where *E. crassipers* is altogether absent second population curve of *Cx. tritaenorhynchus* is much higher (280 PTHC) than the first (38 PTHC) due to extensive breeding of the species in paddy fields leading to fast population build up as soon as rain sets in (Fig. 5.2b). Ground vegetation in adjacent field of *S. bicolor* (Jowar) provides numerous resting places in cool and moist atmosphere for the vector species. Population gradually declines with the onset of winter season and the density of the species reaches as low as 1 PTHC in January.

At Osmanpur, a peculiar behavior was noted. Females of *Cx. tritaeniorhynchus* seem to have developed strong affinity with their day resting sites among water hyacinth. When disturbed they fly only about half an inch and immediately return back, within half a minute, to their microhabitat, which provide favorable temperature, humidity, and shed. This type of hitherto unknown behavior needs to be studied further in order to formulate control measures against these mosquitoes under similar situations in the country.

5.3 Resting Preference of *Culex tritaeniorhynchus* in and Around Delhi

Culex tritaeniorhynchus mosquitoes (males and females) were found to use different types of natural shelters in the form aquatic and land vegetation, seasonal vegetable, and flower available in the area at different period of the year as their

Table 5.2 Shift in outdoor resting habitat of *Cx. tritaeniorhynchus* in and around Delhi

Study villages	Jan–Mar	April–May	June–Oct	Nov–Dec
Nonpaddy area (Delhi)	Carrot, Spinach, Marigold, Brinjal	Water hyacinth, Elephant grass	Water hyacinth, Elephant grass	Brinjal, Hemp
Marshy area (Delhi)	Water hyacinth	Water hyacinth	Water hyacinth	Water hyacinth
Paddy area (Haryana)	Birseem, Marigold	Jowar and bajra	Jowar and bajra	Birseem Marigold

outdoor resting habitat. Seven types of natural shelters were recognized in and around Delhi, viz. *Eichhornia crassipers* (water hyacinth), ground vegetation in *Sorghum bicolor* (jowar) field, *Trifolium alexandrinum* (birseem, a fodder plant), *Calendula officinalis* (merrigold), *Daucus carota* (carrot) *Solanum melanogena* (brinjal), and hemp (wild grass). Of these the most preferred resting habitats were: Jowar (June–October), merrigold (November–December), berseem (December–May) (Table 5.2). Jowar is grown in close proximity of paddy fields. At Osmanpur village, land vegetation included sparsely distributed wild vegetation and abundance of *Cx. tritaenorhynchus* was monitored round the year in perennial aquatic vegetation of water hyacinth marshes that almost encircled the village leaving behind only the elevated approach way from the main road. However, the species was never found to use paddy field as natural day resting habitat.

5.4 First Time Detection of Overwintering Phenomenon in *Culex tritaeniorhynchus* in India

During the month of September, adult mosquito collection employing BPD hop cage method from study areas in and around Delhi revealed that female *Cx. tritaenorhynchus* mosquitoes collected outdoors were in all stages of gonotrophic cycle: unfed, freshly fed, semi gravid, and gravid (Chap. 4: Fig. 4.8). All the three categories of fed population share the same day resting shelters in their preferred vegetation along with the freshly hatched/unfed and male counter parts. This indicates true exophilic behavior of the species in the study area. During November, wild caught *Cx. tritaeniorhynchus* females were predominantly unfed. As night temperature dropped in the area, females stopped taking blood meal and overwinters resting in available local vegetation. Male population totally disappeared for next three cold winter months (December–February). From mid February due to gradual rise in night temperature, overwintering females started taking blood feed. However, during the same period, *Cx. quinquefasciatus* females never found to over winter in the area. In an earlier studies (Das 2003), over wintering phenomenon in *Cx. tritaenorhynchus* females was reported first time from Delhi by the author. This was the first report of over wintering phenomenon in *Cx. tritaenorhynchus* mosquitoes from India.

5.5 Breeding Habit of *Culex tritaeniorhynchus* in and Around Delhi

A variety of larval habitats of *Cx. tritaeniorhynchus* were identified in the study area, viz. paddy fields, temporary pond, road sides burrow pits, and irrigation canal at extensively paddy growing area of Safiabad village, whereas in nonpaddy areas, larval habitats included ground pools, disused irrigation canal, and vast low lying marshy area at Madanpur Khadar and Osmanpur village.

Larval survey conducted at paddy growing area indicated that breeding of *Cx. tritaeniorhynchus* is closely associated with monsoon rains with peak larval abundance observed in the month of August coinciding with increased adult female abundance. Paddy fields were the most favored breeding place followed by burrow pits and ponds (Fig. 5.3a). After paddy was harvested by October end, breeding habitats of the species become limited and confined to burrow pits and temporary pond in November. Thereafter, as the night temperature dropped in the area, females stopped taking blood meal and entered into overwintering phase resting in local vegetation. Breeding of the species totally ceased till mid of February when night temperature gradually rises.

Fig. 5.3 Breeding characteristics of *Cx. tritaeniorhynchus* in and around Delhi. **a** Paddy growing area. **b** Non paddy area

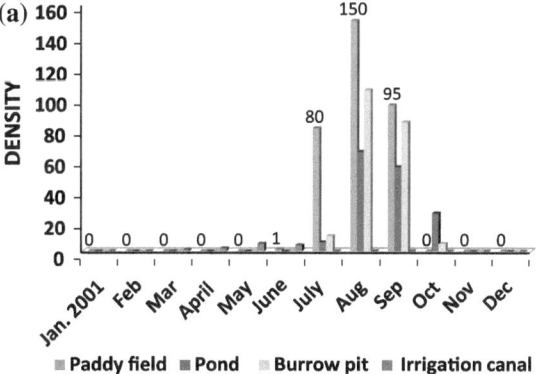

Overwintering females started taking blood feed and compelled to breed on the grassy edges of the irrigation canal. Temporary pond and burrow pits get dried up by January end leaving behind irrigation canal, the only breeding habitat available for the species amid a vast dry area. As soon as monsoon sets in (July), environmental conditions change dramatically in the area causing a shift in breeding habitat of the species from irrigation canal to variety of breeding habitats, viz. paddy fields, temporary ponds, and burrow pits. Per dip density of the species reached its peak (150) within a month (August) due to extensive breeding of the species in paddy fields (Fig. 5.3a).

However, larval survey conducted at nonpaddy area of Delhi revealed a different picture. Ground pools, disused irrigation canal, and vast marshy area with extensive vegetation cover of water hyacinth provide congenial resting habitats for entire population of *Cx. tritaeniorhynchus* including favorable breeding sites for the species round the year (except for 3 months from December, January, and February). The larval population started building up from February onward and reached its peak (85 per dip) during the month of June, thereafter started declining and touched zero during November–January (Fig. 5.3b). In contrast, in paddy growing area larval population started building up from July onward coinciding with paddy transplantation and arrival of monsoon rains in the area. Important reasons recognized behind the difference in breeding pattern of the same species within a distance of 40 km: (i) presence of perennial breeding sites of the species in disused irrigation canal at Madanpur Khadar village and vast marshy area at Osmanpur village, both covered with dense water hyacinth in which ideal microhabitat coupled with "crowding effect" effect of the species was observed during prolonged dry months from March to June and (ii) as soon as monsoon sets in July, additional numerous breeding sites of the species are created in both Madanpur khaddar and Osmanpur village in the form of ground pools in which breeding of the species gets diffused.

5.6 Natural Microbial Organisms Found Infecting JE Vector Larvae in and Around Delhi

5.6.1 Discovery of a New Microbial Control Agent for Mosquito Vectors of Human Diseases

For the first time in science, a ciliated protozoa, *Chilodonella uncinata* (Fig. 5.4) with no previous published record of pathogenicity against mosquitoes, was detected and identified to be causing significant mortality in natural population of JE vector larvae (*Cx. tritaeniorhynchus* and *Cx. pseudovishnui*) in and around Delhi (Das 2003).

5.6.2 Events that Lead to the Invention

During August 2000, an inexperienced hand joined my laboratory which was engaged in JE vector survey. He was given practical demonstration of collecting mosquito larvae form the natural breeding habitats as well the precautions to be followed for transporting them to the laboratory alive. Next day, he brought mosquito larvae from paddy fields but by time they were brought to lab all larvae died. He was told to repeat, however, again 100 % mortality, prompted me to detect the causative organism. Initial investigation only revealed very high mortality in natural population of JE vector larvae (*Cx. tritaeniorhynchus* and *Cx. pseudovishnui*). About one-tenth of the dead larvae were transparent and on examination under the microscope were found to be severely infected with endoparasitic ciliates, which were found moving inside the entire body: hemocoel, head capsule, siphon, saddle, and anal gills (Fig. 5.4a, b).

As there was no expertise available at our institute (NICD), it took several months as well series of follow-up experiments to reveal the identity of these parasites responsible for such high epizootic in natural population of JE vector larvae.

5.6.3 Problems Encountered in Detection, Isolation and Identification of the Causative Organism

A few species of lower invertebrates belonging to the genera viz.: *Paramecium, Euplotes, Stylonychia* (Ciliated Protozoa); Thecate *Amoeba* (Phyllum Protozoa) and a species of Rotifer were recorded many a times from the same natural habitats of mosquito larvae. According to the literature and experts in the field, none of these aforesaid microorganisms have been recorded to be pathogenic to mosquito

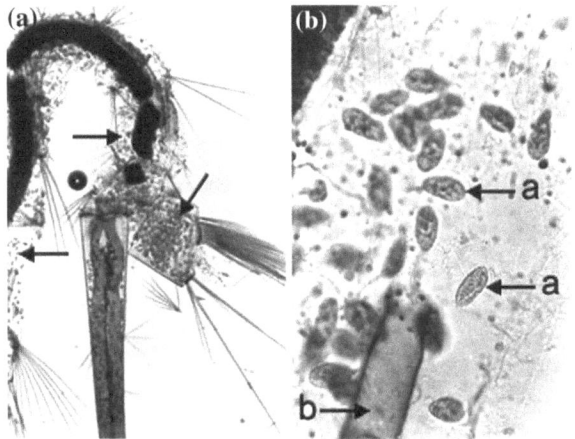

Fig. 5.4 Wild caught transparent dead JE vector larva (*Cx. tritaeniorhynchus*). **a**, Fourth instar larva: *arrow* showing endoparasites (*Ch. uncinata*). **b** Same in higher magnification: a endoparasites. b disintegrated alimentary canal (Adapted from Das 2003)

Fig. 5.5 *Ch.uncinata* (Ehrenberg) 1838 [Adapted from Das (2003)]

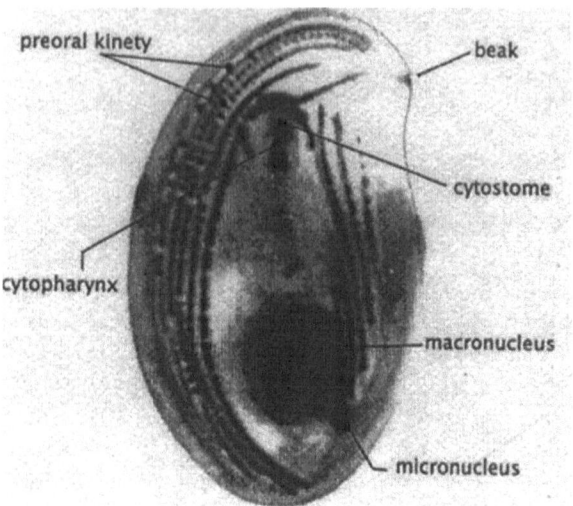

larvae. A species of Vorticella (Protozoa) was found attached to dead mosquito larvae (Vorticellids are not pathogenic to mosquito larvae. However, under special circumstances, when large numbers of Vortecellid get attached to one mosquito larva, with the increased weight the larva is drowned and dead).

These dead mosquito larvae were washed in distilled water and kept in aliquots for a few days under laboratory condition and observed under microscope. Another dorso-ventrally flattened ciliated protozoa not noticed earlier now appeared in the water containing the dead larvae. At that particular point, there was an indication that probably these dorso-ventrally flattened ciliated protozoa might be the causative microbe killing the mosquito larvae. Subsequently, these were identified by the experts in the field as *Chilodonella uncinata* (Ehrenberg), 1838 (Fig. 5.5) (Subphylum: Ciliophora: Cryptophorida: Chilodonellidae) in wet mounts (Foissner 1991). This was for the first time the genus *Chilodonella* was found to be pathogenic to mosquito larvae in nature. As per the available literature, three species of *Chilodonella*, viz. *Ch. cucullulus*, *Ch. rhesus* and *Ch. spiralidontis* (Muller 1773; Ghosh 1929; Bhatia and Mullick 1930) were reported earlier from India (Bhatia 1936). Therefore, the discovery of the present pathogenic ciliate *Ch. uncinata* in and around Delhi was a new record for the species from India.

5.6.4 Biological Characteristics of Chilodonella uncinata

5.6.4.1 Description

Body flat, dorso-ventrally compressed; size 30–50 μm; contractile vacuole two in number; macronucleus, micronucleus; cytopharynx long, and curved.

5.6.4.2 Life Cycle

The organism has a unique mode of behavior. It is a facultative parasite of mosquito larvae and its distribution as a free swimming (trophont) stage is scanty. In presence of susceptible mosquito larvae, it changes itself and become parasitic in habit (Das 2003, 2008). Number of them simultaneously attack one single larva at different points (head, thorax, abdomen, siphon, anal fin, antennae, etc.) and enter body cavity of the host larva by drilling through the host cuticle. Penetration points are permanently marked by circular holes on the cuticle (Fig. 5.6). This type of mode of entry of *Ch. uncinata* from outside through cuticular cyst is similar to that reported in case of two parasitic ciliates, viz. *L. stegomyiae* and *L. clarki* (Lamborn 1921; Narain et al. 1996; Corliss and Coats 1976) and one parasitic fungi (*Coelomomyces sp*) (Wong and Pillai 1980). Of these, *L. clarki,* a free-living ciliated protozoa with many desirable attribute of a potential biocontrol agent (Washburn 1995), is known to parasitise tree hole-breeding mosquito species (*Aedes sirensis*). *Coelomomyces* is known to parasitise Anopheline larvae in paddy fields in South India (Chandrahas and Rajagopalan 1979) and in a temporary pool in North India (Gugnani et al. 1963). However, *Coelomomyces* has a complex life cycle and requires both the primary host (mosquito larvae) and intermediate host (copepod) for reproduction (Lacey and Lacey 1990).

Once inside the body cavity, *Ch. uncinata* start multiplying and increase their number exponentially at the expense of the internal viscera of host larva, ultimately host larva dies. At this stage the larva is buff colored and even after the death of host larva, the parasites continue to multiply for some more time till the nutrients inside the host body get exhausted. By then the host larva turns transparent and several hundred parasite come out of the carcasses of dead larvae to attack

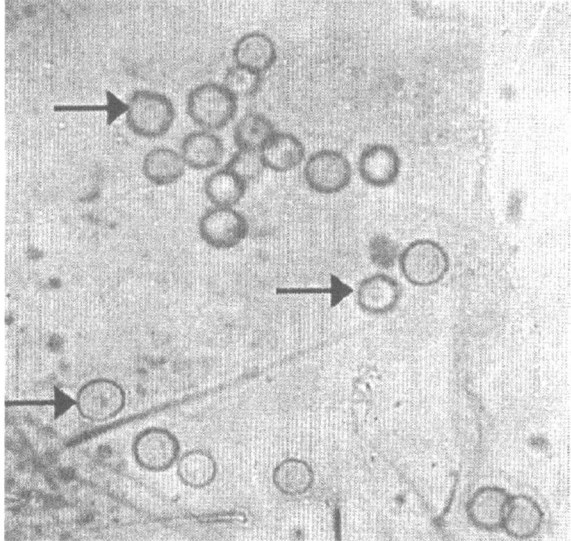

Fig. 5.6 Entry holes of parasite *Ch. uncinata* on the dead host cuticle (*Cx. tritaeniorhynchus*) [Adapted from (Das 2003)]

fresh susceptible host larva to continue the cycle (recycle in the environment). The organism is so virulent that even a few of them cause chronic infection leading to death in susceptible host mosquito larvae (Das 2004). Similar behavior was observed elsewhere with another ciliate parasite *Lambornella clarki* on its mosquito host *Aedes sirensis* (Washburn 1995).

5.6.4.3 Transovarian Transmission

An attempt was made to test the transovarian transmission potential of *Ch. uncinata* through its mosquito host. During August 2000, 25 fed *Cx. tritaeniorhynchus* were collected from Safiabad village (Sonipat District: Haryana) having extensive paddy cultivation and were kept for egg laying separately in glass specimen tubes. Water in experimental tubes was examined using 100X magnification for the presence candidate pathogen for a period of 7 days. Within a period of 2–5 days *Ch. uncinata* was detected in water of 80 % of specimen tubes used for egg laying of wild caught *Cx tritaeniorhynchus* confirming the transovarian transmission potential of *Ch. uncinata* through its mosquito host. Eggs were laid in 55 % and larvae were hatched in 30 % of the experimental tubes, respectively. However, all the larvae hatched died before reaching to 4th stage and subsequently all of them were found to be infected with *Ch. uncinata* (Das 2003).

5.6.4.4 Tolerant to Dryness

In order to determine the effect of dryness on the stability of the pathogen a number of plastic cups containing *Chilodonella uncinata* NICDENTBPL 13106 (North India strain) were allowed to dry at room temperature, re-flooded with distilled water after a varying period of time and examined for revival of the pathogen using 100X magnification. *Ch. uncinata* reappears in the dry plastic cups in a span of 2–5 days' time indicating thereby its capability to stand desiccation (Das 2003).

5.6.4.5 Blood Feeding Drive Inhibition of *Cx. tritaeniorhynchus* Induced by Parasite *Ch. uncinata*

An attempt was made to establish parasite-induced inhibition of blood feeding drive of host mosquito species. Wild caught *Cx. tritaeniorhynchus* mosquitoes were collected during August 2001) when abundance of both the parasite and its host (*Cx. tritaeniorhynchus*) is at the peak from two areas: Safiabad village (with very high influence of the parasite) and Madanpur Khaddar (with negligible influence of *Ch. uncinata*). 25 unfed females *Cx. tritaeniorhynchus* each collected from both the areas were kept in separate $12'' \times 12'' \times 12''$ cloth cage and pigeon bait was provided as blood feed. Mosquitoes in both sets of the experiment were provided fresh feed every night. While 85 % of female mosquitoes from Madanpur

Khaddar were found fed after 2nd night, only 10 % of those females from Safiabad village had taken blood by 6th night confirming *Cx. tritaeniorhynchus* mosquitoes infected with *Ch. uncinata* were significantly less responsive toward a vertebrate host as compared to uninfected females. Similar blood feeding drive inhibition by tree hole breeding *Lambornella clarki* infected *Aedes sirensis* was demonstrated elsewhere (Yee 1995).

5.6.5 Role of Ch. uncinata on the Population Dynamics of Cx. tritaeniorhynchus Mosquito in Nature

During JE transmission season in Northern India, population abundance of *Cx. tritaeniorhynchus* at Safiabad village (Non-endemic area) with high prevalence of ciliate parasite (*Ch. uncinata*) was found to be higher as against significantly low population abundance of the host mosquito species recorded from two JE endemic districts, viz. Karnal district (Haryana), and Saharanpur district (U.P.) where these parasites (*Ch. uncinata*) are either inexistence or poorly represented (Fig. 5.7).

In nature, during August–September abundance, both the parasite and its host (*Cx. tritaeniorhynchus*) are at their peak. Sizeable portion of the parasite (*Ch. uncinata)* infected larvae of *Cx. tritaeniorhynchus* die before they reach to pupal stage; those larvae survive and became pupae: many die; those pupae survive and become adult: carry these parasite in the adult stage. Some of these females were significantly less responsive toward host seeking drive than an uninfected female and only a few succeeded to have blood feed, reach to gravid stage: a part of these infected gravid female mosquitoes failed to lay eggs as the parasite invade the female reproductive organ of the host that remained underdeveloped, while some (infected gravid females) deposit both eggs and parasites (*Ch. uncinata*) in aquatic habitats (Fig. 5.8). Similar observation was recorded by earlier investigators with another ciliate parasites *L. clarki* and its mosquito host *Aedes sirrensis* (Corliss and Coats 1976; Egerter et al. 1986; Washburn 1995; Yee 1995). The parasites invade the ovaries, resulting in females that are castrated and that generally cannot produce eggs (Hawley 1985; Egerter et al. 1986).

By September end paddy fields are kept dry and by middle October paddy is harvested, winter sets in, females stop taking blood feed and enters into overwintering phase resting in local vegetation till the time environment warms up by mid February. Thus, at peak mosquito abundance period (August) in Safiabad village, majority of *Cx. tritaeniorhynchus* mosquitoes are young females that are not capable of reaching the infective stage. So there is no scope of disease (JE) transmission during peak transmission period (October) in places where these parasites (*Ch. uncinata*) have natural check on JE vector species as is seen in Safiabad, one of the study area in the present study. It may be mentioned that other parameters for an impending outbreak of JE remained the same with availability of water birds (pond herons) and pigs were found in plenty. Thus, it is likely that *Ch. uncinata*

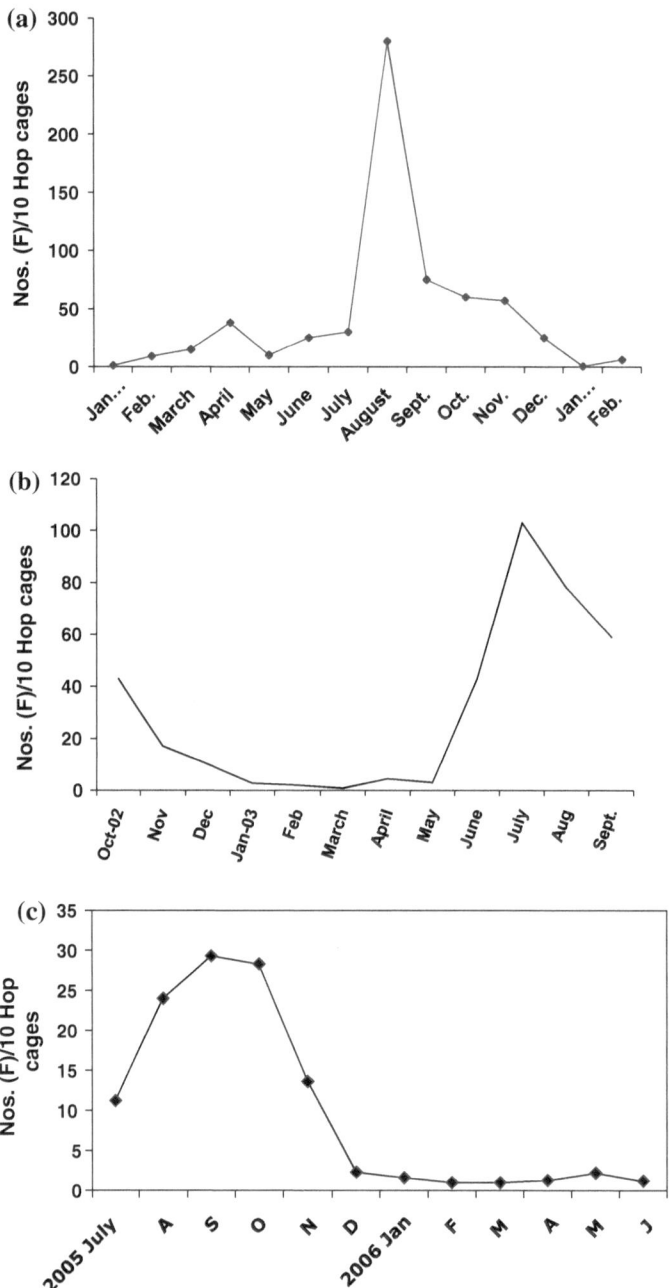

Fig. 5.7 Population abundance of *Cx. tritaeniorhynchus* (hop cage method) in endemic and Nonendemic areas of Northern India. **a** Sonipat District (Non-endemic). **b** Karnal District (Endemic). **c** Saharanpur District (Endemic)

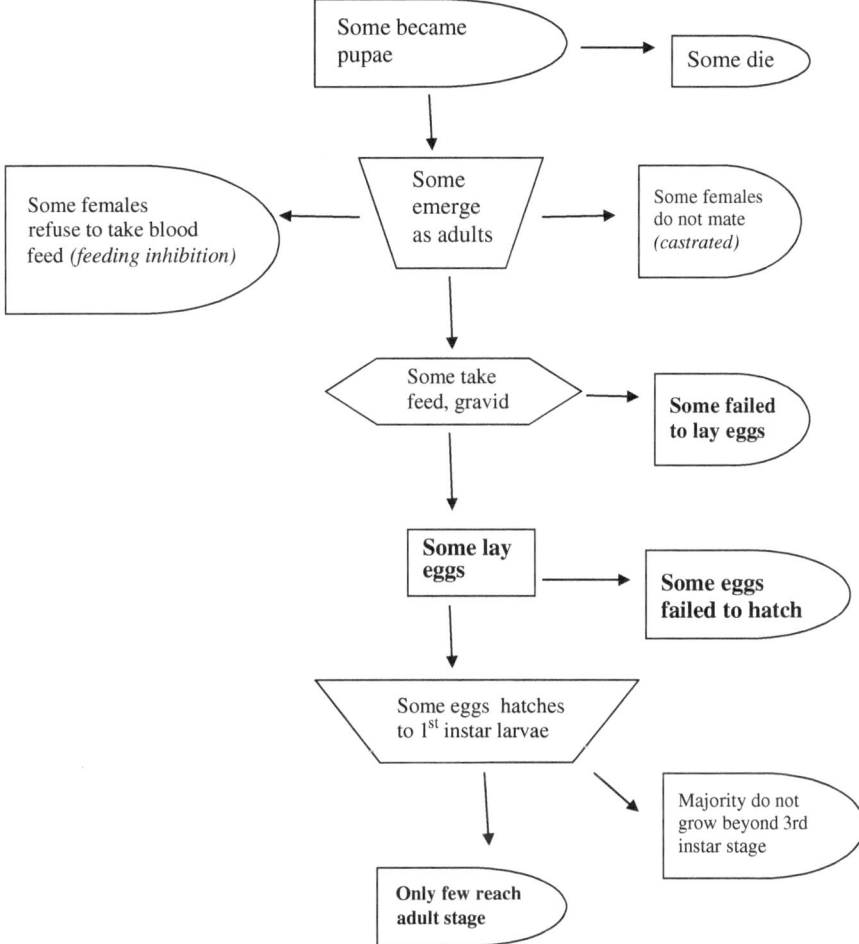

Fig. 5.8 Role of *Ch. uncinata* on population dynamics of *Cx. tritaeniorhynchus* in nature as observed in Sonipat District, a non endemic area in Northern India

induce natural check on the abundance of infective JE vector population at Safiabad village with majority have developed inhibition in taking blood feed and the area so far remained free from JE.

The situation in other two study village located in Delhi, viz. Madanpur Khaddar and Osmanpur were entirely different with extensive water hyacinth problem where impact of *Ch. uncinata* on JE vector population was observed to be negligible. Irrigation canals, ponds, and marshy areas with prolific growth of floating and emergent vegetation predominantly of water hyacinth contribute immensely to the adult mosquito population of *Cx. tritaeniorhynchus*. Analysis of data from these habitats in and around Delhi (Table 5.3) indicates that *Cx.*

Table 5.3 *Chilodonella uncinata* infection detected in breeding places of *Culex tritaeniorhynchus* mosquitoes from study areas of Delhi and Haryana

Study area	Type of habitat	*Chilodonella uncinata* infection in *Culex tritaeniorhynchus* larvae	
		Nos. found positive/ Nos. examined	% Mortality
Safiabad (Haryana)	Paddy field	2174/2703	80.43
	Wells	86/97	88.66
	Burrow pits	797/895	83.91
Madanpur Khadar (Delhi)	Irrigation canal, Water hyacinth pool	3/121	2.47
Osmanpur (Delhi)	Water hyacinth marshes	2/72	2.77

tritaeniorhynchus has a comparatively low level of susceptibility to *Ch. uncinata* infection (mortality rate >3.0) as compared to other habitats, viz. seasonal ponds, paddy fields, burrow pits, and wells in Safiabad having aquatic vegetation excluding water hyacinth (mortality rate ranging from 80.43 to 88.6). Though the reason for reduced impact of the parasite under the influence of water hyacinth was not fully understood but it appears darkness due to entangled leaves of hyacinth plant that blocks sunlight from reaching the water underneath might be one of the limiting factors supported by the laboratory evidences, viz. (i) significant decrease of cell growth of *Ch. uncinata* when tried to colonize under dark laboratory condition, (ii) under the light of microscope the organism was found to be very active.

Both these areas along with similar areas of Delhi with large number of vagabond pig population and water birds are at risk areas as far as outbreak of JE is concerned. Water hyacinth marshes of irrigation canal including those on the sides of Agra canal and ponds provide ideal breeding as well resting ground of these mosquitoes. During rainy season in the absence of cattle population JE vector population increases in the focal areas. In the event of increased movement of infected pig population for slaughter purposes from nearby JE endemic areas of Haryana poses a threat to an impending outbreak of JE in these areas. The apprehension became a reality when JE cases were reported from north and east Delhi with 14 people were found positive for JE virus during 2011.

5.6.6 Colonization of Ch. uncinata Under Laboratory Condition

These organisms were isolated from infected adult mosquitoes and wild caught diseased mosquito larvae. They were colonized under laboratory condition on artificial media using simple technique in glass container and filtered through 10 μm mesh cloth. The culture filtrate was the first strain developed and designated as

Chilodonella uncinata NICDENTBPL 13106 (North India strain). *Ch. uncinata* could be produced on large scale under laboratory condition (Temp. 27 ± 2 °C), Humidity (70–80 %) within 48–72 h. *Ch. uncinata* NICDENTBPL 13106 (South India strain) was developed using the parasites laid by infected *Cx. tritaeniorhynchus* mosquitoes brought from a mosquito colony of an institute in south India. These parasites (*Ch. uncinata*) from south were dangerously virulent and decade old *Anopheles stephensi* colony of NICD got badly destroyed by accidental contamination with these parasites. The South India strain was found to be very effective when tested at VCRC, Pondicherry against second instar larvae of *An. stephensi, Culex quinquefasciatus,* and *Aedes aegypti* obtained from the cyclic colonies of the institute. Comparatively, higher doses were required when North India strain was used indicating thereby difference in virulence depending upon geographical strain in otherwise morphologically similar organism (Das 2008).

Subsequently, culture procedure was modified thereby making it more efficient and standardized using the facilities available at the Department of Biosciences, Jamia Millia Islamia (JMI), a Central University, Delhi was designated as *Chilodonella uncinata* BP 610. The strain was deposited with the International depository authority (IDA)—ATCC, U.S.A. for which accession number: ATCC PRA-373 was allotted.

5.6.7 Formulation

Scmi-dry formulations were prepared using both the concentrated culture strain, viz. *Ch. uncinata* NICDENTBPL 13106 (North India strain) and *Ch. uncinata* BP 610 (an improved North India strain) with base material as matrix to entrap the active ingredient under laboratory conditions. In the formulation, the active ingredient—the protozoan parasite (*Ch. uncinata*) remains inactivated, when released in water works as an effective biological control agent for mosquito vectors of human diseases (Malaria, Japanese encephalitis, and Dengue/Chikunguinia), the parasites get reconstituted into actively free swimming (trophont) form after a gap of 2 h to few days. *An. stephensi* larvae were found to be more susceptible to protozoan formulation made out of NICDENTBPL 13106 (Das 2008) and *Ch. uncinata* BP 610.

While the formulation using culture strain of *Ch. uncinata* NICDENTBPL 13106 (North India strain) was unable to withstand extreme cold (below >4 °C) climatic situation in Delhi, *Ch. uncinata* BP 610 formulation prepared with modified culture strain of *Ch. uncinata* BP 610 is resistant to extreme cold condition in the laboratory condition at JMI with excellent recovery rate even after 18th months.

5.6.7.1 Advantage of *Ch. uncinata* Formulation Developed

Because of the extreme vastness of paddy fields as larval habitat of JE vectors in developing countries and the limitations of resources in man power and material required for larval control, a microbial control agent will not only have to be efficacious against target mosquitoes and safe to nontarget organisms; but it should also replicate, persist, and disperse within the environment (Lacey and Lacey 1990).

The active ingredient of formulations, viz. *Ch. uncinata* NICDENTBPL 13106 and *Ch. uncinata* BP 610 developed at NICD and JMI, respectively, has many properties of a good biological control agent. These are: (i) easy to colonize under laboratory condition, (ii) can be produced in large scale using simple technology, (iii) tolerant to desiccation, (iv) robust and not sensitive to vagaries of agricultural pesticides, (v) facility to recycle in the environment, (vi) can pass from infected adult mosquito to offspring, (vii) not harmful to larvivorous fish, (viii) female mosquitoes infected with *Ch. uncinata* are significantly less responsive toward a vertebrate host as compared to uninfected females.

The formulation when released in water works as an effective biological control agent for mosquito vectors of human diseases (Malaria, Japanese encephalitis, and Dengue/Chickunguinia), *Ch. uncinata* gets reconstituted, attacks mosquito larvae to enter into the hemocoelomic cavity of the larvae, multiplies and releases numerous minute motile spores, ultimately kills the host larvae to finally escape the cadaver of host larva.

5.6.8 Mode of Entry of Ch. uncinata Inside the Host Body and Subsequent Histopathology

During July 2005, in order to know the mode of entry *Ch. uncinata* and subsequent histopathology of host mosquito larva, a simple experiment was undertaken at VCRC under the guidance of Dr. K. Balaraman and Dr. P. Jambulingam. The protozoan, *Ch. uncinata* NICDENTBPL 13106 (South India strain) was obtained from the culture collection of NICD. Second instar larvae of *An. stephensi* were obtained from the cyclic colonies of VCRC. The *Ch. uncinata* cells were counted as per numbers present per ml of culture used and tested for infection and mortality of mosquito larvae. For this, 25 s instar larvae held in a disposable cup with 125 ml chlorine free tap water added 1 ml of wide range dose (80,000 cells/ml) of *Ch. uncinata* culture and observation was made at hourly interval. No larval food was provided to avoid interference with the experiment. On examination under the microscope, the larva was found very irritable as *Ch. uncinata* was found to attack at several points after a few hours of post treatment and very few trophont stage of *Ch. uncinata* were visible outside in the cup. After 5 h of post treatment a larva was slightly teased, instantly numerous endoparasitic stage of *Ch. uncinata* were found oozing out from the teased larva (Fig. 5.9a). After 48 h all the larvae were dead, almost transparent and endoparasitic stage of *Ch. uncinata* were found moving inside the entire host body (Fig. 5.9b). Entry points (shown by arrows) of the parasite on the cuticle of host body were also visible (Fig. 5.9c). Finally, after few days, trophont stage of the parasite increased outside the larvae.

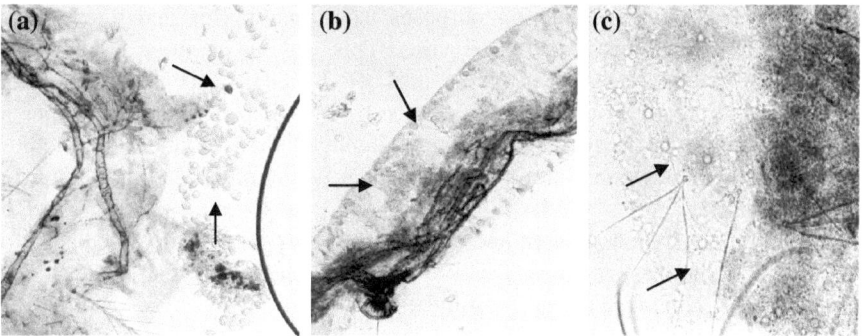

Fig. 5.9 Mode of entry of *Ch. uncinata* inside host (*An. stephensi*) body. **a** Teased larva after 5 h of post treatment (*arrows* showing endoparasites). **b** Another larva after a few days of post treatment. **c** Entry points of parasites on host body

After the above stages were demonstrated, it became clear to understand the events that take place in the life history of otherwise free-swimming ciliate *Ch. uncinata* in presence of a susceptible host in the breeding habitat. This type of mode of entry of microbe from outside is known only in two cases namely. *Lambornella* (another ciliated protozoan) and *Coelomomyces* (parasitic fungi) of mosquito larvae.

5.6.9 Filling of National and International Patents

National and international (in seven countries) patent applications were filed on "Microbial control agent for mosquito vectors of Human diseases" (Inventor: Dr. Bina Pani Das; Co-Applicants: Department of Biotechnology (DBT), Ministry of Science & Technology & National Institute of Communicable Diseases) supported by DBT in 2001. So far, patent has been granted by six countries, viz.: Bangladesh (Patent # 1003897, dated 08.08.2005); U.S.A. (Patent # 7141245, dated 28.11.2006); Australia (Patent # 2002217423, dated 28.06.2007); Sri Lanka (Patent # 13134, dated 11.12.2007); Vietnam (Patent # 6774, dated 31.12.2007); Philippines (Patent # 1-2003-500738, dated 01.08.2008).

5.7 Situation Specific Vector Control Measures in and Around Delhi

Though these areas are not endemic for JE, during October 2011, sporadic human cases of JE were reported almost simultaneously from several areas of Delhi. These were preceded by reporting of large number pigs that were probably brought to Delhi for slaughtering purposes from neighboring endemic state of Haryana were found positive for JE virus. This was an alarming situation for Health Authorities responsible for control of Japanese encephalitis. Control of JE vectors requires coordinated efforts by Municipal/Urban Development Administration and Health authorities including Indian Council of Medical Research, National Vector Borne Disease Control Programme, National Centre for Disease Control.

Since JE vector bionomics in and around Delhi has been clearly established it has become rather easy to suggest JE vector control/management strategies in and around Delhi. Among culicine mosquitoes, *Cx. tritaeniorhynchus,* principal JE vector species were found to be the predominant species in and adjoining areas of Delhi.

5.7.1 Control Strategies in Delhi

Vast marshy areas located particularly in East and south east Delhi provide both extensive breeding ground as well as primary day resting sites for *Cx. tritaeniorhynchus* mosquitoes round the year. The vector species is zoophilic but in the absence cattle population from major part of Delhi even moderate abundance of vector species may lead to transmission during monsoon months as adequate number of vagabond pigs with pond herons and cattle egrets were found in plenty in the area. Therefore, controlling water hyacinth problem in Delhi is the big step forward in keeping Delhi free from JE cases in future.

Water hyacinth originally came from South America where these plants are kept in check by natural enemies. But in new environments like India, Africa, etc., these natural enemies are missing so the plants grow wildly, double every 5–15 days. It grows and grows until it covers the water with a thick floating mat of tangled weed. This causes terrible problems for people: (i) The weed provides numerous ideal day resting sites for disease causing mosquitoes like *Cx. tritaeniorhynchus* round the year, and the water underneath the plant is the ideal breeding ground of the mosquito species, (ii) Dangerous animals like snakes also hide amongst the weeds, (iii) These plants use up precious water. Water is lost over three times faster than from clear water surface because of transpiration from the leaves (ITDG, Technical brief on water hyacinth).

5.7.1.1 Control of Water Hyacinth

Co-ordinated efforts of Municipal Corporation of Delhi, Irrigation Department and Health Authority of Delhi Government are required to remove water hyacinth from aquatic habitats located in Delhi. Water hyacinth can be controlled using three methods: (i) Chemical Control, (ii) Physical Control, and (ii) Biological control. Small infestation of water hyacinth can be successfully controlled by application of herbicides. However, use herbicide leads to environmental and health hazard, especially where people collect water for drinking and washing purposes.

Physical Control by mechanical removal of water hyacinth is the best short term solution. It is, however, costly and requires the use of both land and water vehicles in places where the infested area includes deep water body. But in most cases in Delhi, water hyacinth problem includes vast shallow marshes, where physical removal is comparatively easier. As labor force is economically feasible in our country, coordinated attempts may be made by municipal as well as irrigation Department in consultation with health authorities for a long-term solution of water hyacinth menace in Delhi and remove the ideal breeding ground for *Cx. tritaeniorhynchus*, the local JE vector in the area. Fifty years back Bangladesh (Erstwhile East Pakistan) was able to control mosquito menace adequately by simply physical control of water hyacinth in the country. Apart from health benefit, both biogas and fertiliser can be generated by utilization of water hyacinth after its removal from the water body. Sankar Ganesh et al. (2005) used three phase volatile fatty acids (VFAs) extraction-biogas-generation-vermicompositing system for water hyacinth utilization. The system fully utilizes the weed partly in generating energy (in the form of biogas) and party in producing fertilizer in the form of vermicompost (Abbasi et al. 2012). Following the process total disposal of water hyacinth is possible by partly generating biogas (energy) and partly producing fertilizers in the form of vermicompost.

Since 1970s biological control agents with three species of weevils, viz. *Neochetina bruchi, N. eichhorniae* (known to feed on the plant) and *Sameodes albiguttalis*, the water hyacinth borer are released to deal with water hyacinth problem. Although meeting with limited success, the weevils have since been released in more than 20 countries (ITDG, Technical brief on water hyacinth).

5.7.1.2 JE Vector Management

Using mass media, public around 5 km of an extensive marshy area, irrigation canal, and pond infested with profuse growth of water hyacinth should be advised: (i) to use of door and window screen, (ii) repellents, and (ii) burning of herbal materials like dry neem leaves/dry coconut husks with resin in an incense pot during evening hours (1 h before and after dusk) for 4 months (July–October) to minimize man–vector contact.

5.7.2 Control Strategies in Paddy Growing Areas of Sonipat District (Haryana)

Since parasite *Ch. uncinata* was found to have adequate natural check on JE vector breeding in extensive paddy growing areas in Sonipat district of Haryana, no chemical and biological control measures are required to control JE vector species in the area till the time similar situation continues. However, it was noticed, during JE vector survey carried out at village Safiabad in August–September 2011, that some of the local farmers had switched on to other crop practices instead of paddy cultivation due to shortage of power supply required to run the tube wells for irrigating paddy fields. Therefore, the District Authorities need to take appropriate measures to ensure requisite power supply for irrigation purposes and have the advantage of natural check on JE vector population in the area. During peak mosquito abundance in monsoon months, in order to reduce mosquito nuisance as well prevent man–mosquito contact general public in area need to be educated to use repellents and burn dry neem leave during evening hours to repel mosquitoes.

References

Abbasi T, Abbasi SM, Abbasi SA (2012) Biogas Energy. Springer Briefs in Environmental Science. New York, I-vii+ 1–169

Bhatia BL (1936) Protozoa: chiliophora. The fauna of British India. Taylor and Francis, London

Bhatia BL, Mullick BK (1930) On some fresh water ciliates from Kashmir. Arch Protistenk 72:390–403

Chandrahas RK, Rajagopalan PK (1979) Mosquito breeding and the natural parasitism of larvae by a fungus, *Coelomomyces* and a mermithrid nematode, *Romanomermis*, in paddy fields in Pondicherry. Indian J Med Res 69:63–70

Corliss JO, Coats DW (1976) A new cuticular cyst producing tetrahymenid ciliate, *Lambornella clarki* n. sp. and the current status of ciliatosis in culicine mosquitoes. Trans Am Microsc Soc 95:729–739

Das BP (2000) A new technique for sampling outdoor resting population of *Culex tritaeniorhynchus*, Vevtor of Japanese encephalitis. Fourteenth National Congress of Parasitology, New Delhi, 23–26 April, 2000, Abstr. No. PS-15, PP 133–134

Das BP (2003) *Chilodonella uncinata*—a protozoa pathogenic to mosquito larvae. Curr Sci 85:483–489

Das BP (2004) Process for preparation of a microbial control agent. Pub. No. US 2004/0219692, United States Patent Application Publication

Das BP (2008) New microbial insecticide—a discovery by accident. Invent Intell 43:26–28

Das BP (2009) BPD hop cage method—a new device of collecting mosquitoes for effective JE vector surveillance. Invent Intell 44:24–25

Das BP, Rajagopal R, Akiyama J (1990) Pictorial key to the species of Indian Anopheline mosquitoes. J Pure Appl Zool 2:131–162

Egerter DE, Anderson JR, Washburn JO (1986) Dispersal of parasitic ciliate *Lambornella clarki*: implication for ciliates in the biological control of mosquitoes. Proc Natl Acad Sci U S A 83:7335–7339

Foissner W (1991) Basic light and scanning electron microscopic methods for taxonomic studies of ciliated protozoa. Eur J Protistol 27:313–330

Ghosh E (1929) A new ciliate from the intestine of the common Bengal monkey (*Mucacus rhesus*). J R Microsc Soc 15–16

Gugnani HC, Wattal BL, Kalra NL (1963) A note on *Coelomomyces* infection in mosquito larvae. Bull Natl Soc India Mal Mosq Borne Dis 2:333–337

Hawley WA (1985) Population dyanamics of *Aedes sirensis*. In: Lounibos LP, Rey JR, Frank JH (eds.) Ecology of mosquitoes, Proceedings of a workshop. Florida Medical Entomology Laboratory, Vero Beach, Florida, pp 167–184

ITDG Technical Brief on Water Hyacinth Control and Possible Uses www.itdg.org/docs/technical_information_service/water_hyacinth_control.pdf

Lacey LA, Lacey CM (1990) The medical importance of rice land mosquitoes and their control using alternatives to chemical insecticides. J Am Mosq Control Assoc (Suppl) 2:1–93

Lamborn WA (1921) A protozoan pathogenic to mosquito larvae. Parasitology 13:213–215

Menon PKB, Rajagopalan PK (1976) A note on *Culex tritaeniorhynchus* Giles, 1901, in villages near Delhi. Indian J Med Res 64:709–712

Muller OF (1773) Verminum terrest, et fluviatil s. animal. In-fusor. etc. historia. Hafniae et Lipsiae, Parts I & II

Narain K, Prakash A, Bhattacharya DR, Mohanta J (1996) Endoparasitic Hymenostome ciliate, a potential biocontrol agent naturally infecting Anopheline larvae in Assam—a preliminary report. J Commun Dis 28:139–142

Rahman SJ, Wattal BL, Bhatnagar VN (1978) *Culex tritaeniorhynchus* Giles in village Arthala, Uttar Pradesh (India) and possible methods of its control. J Entomol Res 2:79–87

Reuben R (1971) Studies on the mosquitoes of North Arcot District, Madras state, India. Part 1. seasonal densities. J Med Entomol 8:119–126

Reuben R, Tewari SC, Hiriyan J, Akiyama J (1994) Illustrated key to genera of Culex (*Culex*) associated with Japanese encephalitis in Southeast Asia (Diptera: Culicidae). Mosq Syst 26:75–96

Sankar Ganesh P, Ramasamy EV, Gajalakshmi S, Abbasi SA (2005) Extraction of volatile fatty acids (VFAs) from water hyacinth using inexpensive contraptions, and the use of the VFAs as feed supplement in conventional biogas digesters with concomitant final disposal of water hyacinth as vermicompost. Biochem Eng J 27(1):17–23

Sirivanakarn S (1976) Medical Entomology studies III. A revision of the subgenus *Culecx* in the oriental region (Diptera: Culicidae). Contrib Am Entomol Inst (Ann Arbor) 12(2):1–272

Washburn JO (1995) Regulatory factors affecting larval mosquito populations in container pool habitats: implication for biological control. J Am Mosq Control Assoc 11:279–283

Wong TL, Pillai JS (1980) *Coelomomyces opifexi* Pillai and Smith (Coelomomycetacae: Blastocladiales). VI. Observation on mode of entry into *Aedes australis* larvae. N Z J Zool 7:135–139

Yee WI. (1995) Behaviors associated with egg and parasite deposition by gravid and *Lambornella clarki*-infected *Aedes sirensis*. J Parasitol 81(5):694–697

Chapter 6
Ecology of *Culex tritaeniorhynchus* in Karnal District (JE Endemic Area), Haryana State in Northern India

Abstract Japanese encephalitis, a mosquito-borne viral disease has been a serious public health problem in Karnal District, Haryana since 1990. The district remained endemic for JE affecting children with very high case fatality rate ranging from 45 to 100 %. In spite of the seriousness of the disease, no published record was available regarding abundance of JE vector species, the entomological evidence of the disease from anywhere in the state. This chapter presents the results of ecological studies undertaken on *Culex tritaeniorhynchus*, principal JE vector species in selected study sites in Karnal District. The seasonal abundance of vector species closely follows rainfall pattern with peak vector abundance occurred during peak monsoon period in the area. In Karnal District, *Cx. tritaeniorhynchus* breeds principally in paddy fields that remained water logged for nearly 3 months (July–September). JE virus infection was detected by ELISA in one mosquito pool comprising of female *Cx. tritaeniorhynchus* mosquitoes collected from Peri-urban area of Karnal town 10 days before the appearance of human cases in the area. Selective immunization of children below >15 years and integrated vector control strategies are suggested to contain child death due to Japanese encephalitis in Karnal District.

6.1 Introduction

Karnal District falls in the north-east part of the Haryana State and lies between North latitudes $29°25'05''$ and $29°59'20''$ and East longitudes $76°27'40''$ and $77°13'08''$. The district is bordered by the river Yamuna in the east, Panipat district in the south, Kaithal district in the west, Kurukshetra, and Yamunanagar district in the north (Fig. 6.1). The district is well connected by roads and railways. The National Highway (NH No.1) runs through the entire length of the district. A broad gage railway line connecting Delhi with Ambala runs almost parallel to the NH No.1. The main townships are Karnal, Indri, Assandh, Nissang, Nilokheri, and Gharaunda. Karnal is the district headquarters. The district has

Fig. 6.1 Study sites (❤)
in Karnal district, Haryana
(2002–2003)

six administrative blocks, viz.: Gharaunda, Indri, Karnal, Nilokheri, Nissang, and
Assandh. There are 26 PHC's, 5 CHC's, and one district Hospital. Karnal District
is one of the most densely populated districts of the state with six municipalities,
434 villages and 380 Gram Panchayats. The total population of the district as per
2001 census is 12,74,183. The river Yamuna provides the major drainage in the
area. Irrigation in the district is done by surface water as well as ground water.
Mostly, 70 % of the net irrigated area is covered through ground water. The area
constitutes almost alluvial plain and forms a part of the vast Indo-Gangetic plain.

The climate of the district is characterized by hot summer and a cold winter. The year may be divided into four seasons. The cold season starts by late
November and extends to the middle of March. It is followed by hot season which
continues to the end of June when the southwest monsoon arrives over the district.
July to September is the southwest monsoon season. After November night temperature decreases rapidly till January which is the coldest month (Fig. 6.2). The
normal annual rainfall of the district is 696 mm, about 82.39 % of the annual rainfall is recorded during June–September.

The main crops in the area are sugarcane (*Saccharum officinarum* Linn.),
paddy (*Oryza sativa* Linn.), wheat (*Triticum aestivum* Linn.), and jowar (*Sorghum
bicolor* Linn.) while minor crops include mustard *Brasscia juncea* (Linn.) and berseem (*Trifolium alexandrinum* Linn.). Broadly, paddy and jowar (millet) are grown
in close proximity during monsoon and post monsoon months (June–October).
Wheat and berseem (a fodder plant) are grown from December to following April
while mustard crop is grown for 2 months (November and December). The millet
and mustard crops support extensive low, ground grassy vegetation where mosquitoes prefer to rest during the day. The fodder plant is low level ground vegetation
and provides resting habitats to *Cx. tritaeniorhynchus* and *Cx. quinquefasciatus*
mosquitoes.

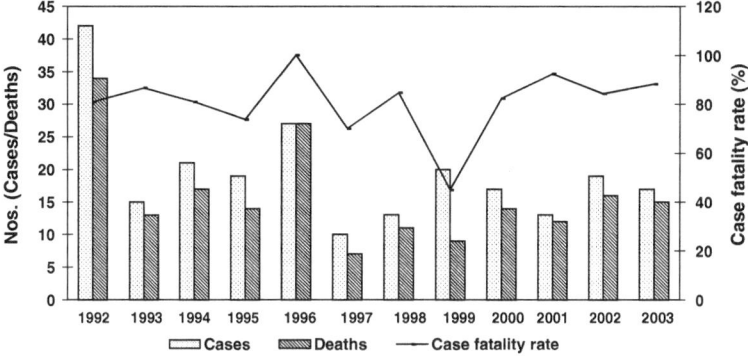

Fig. 6.2   Number of JE cases/deaths, case fatality rate in Karnal district, Haryana (1992–2003) (*Source* DMO, Karnal)

6.2 JE in Haryana State/Karnal District

The activity of Japanese encephalitis virus (JEV) was first time recognized in Haryana in 1990 from nine districts, viz. Karnal, Kurukshetra, Kaithal, Panipat, Rohtak, Sonipat, Jind, Ambala, and Bhiwani. The 1990 epidemic of JE in Haryana reported 294 cases with 205 deaths. Out of the 294 total cases, 229 cases (77.9 %) were reported from four adjoining districts, viz. Karnal, Kurukshetra, Kaithal, and Panipat (Sharma and Panwar 1991). The diagnosis of the outbreak was based on clinical symptoms, detection of JE antibodies in pigs, IgM antibodies in acute human sera samples, and the presence of the vector *Culex vishnui* Subgroup of mosquitoes (Sharma et al. 1991; Prasad et al. 1993). Of these four districts, worst affected was Karnal District which remained endemic for JE affecting children below 15 years (Kar, Saxena 1998) with very high case fatality rate raging from 45 to 100 % (Fig. 6.2). During 2002, first suspected case of JE was reported in second week of October. More number (5) of cases was reported in 2nd and 3rd week of October followed by a decrease of cases as well as deaths in November (Fig. 6.3).

6.3 Ecological Study of JE Vectors in Karnal District

Though Karnal District is endemic for JE since 1990, nothing was known regarding the entomological evidence of the disease from anywhere of the state till the present study that was started in October 2002. Therefore, it became all the more necessary to undertake an ecological study on JE vectors from the state. Because of JE endemicity and high mosquito-genic potential of Karnal District, it was selected for: (i) Evaluation of BPD hop cage method against the existing sampling tools, viz. Drop net and Hand catch method (Chap. 4), (ii) Ecological study on vectors of JEV from JE endemic area of Karnal District in Northern India (present chapter), and (iii) to suggest appropriate integrated methods of vector control. The district is situated within

Fig. 6.3 Weekly trend of
suspected JE cases/deaths in
district Karnal (2002)

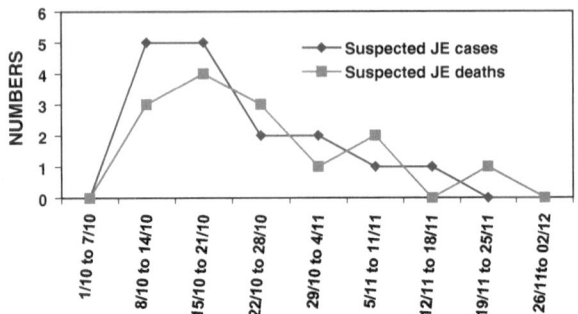

the command area of Western Yammuna Canal, hence was previously endemic for malaria and later on became receptive for JE with intensive rice cultivation supporting extensive JE vector breeding, presence of Ardied birds (cattle egrets, pond herons, the reservoir host), and moderately high pig population (the amplifier host of the virus).

As the new tool (BPD hop cage method) was already proved to be a simple operationally feasible sampling tool for *Culex tritaeniorhynchus* in nonendemic area (Chap. 5), it was used along with conventional hand catch method to study the ecology of JEV transmitting mosquitoes including their virus infection rates in selected areas of Karnal District. Five study sites (four villages and one Peri-urban area in Karnal town) were selected for ecological studies on JE vectors from October 2002 to March 2004. These were: Kutail Kalan (PHC: Kutail), Kambopura and Tikri Kalan (PHC: Indri), Kharkhali (PHC: Madhuban), and National Dairy Research Institute (NDRI), Karnal Town.

Monthly visits were made to each village and adult mosquitoes were collected during 09:00–13:00 h from outdoor resting habitats in jowar, mustard, and berseem crops using BPD hop cage method. Adult mosquitoes were also collected during 08:00–10:00 h from domestic animal shelters and inside houses including living room, store room with help of sucking tube, and torch light. Animal shelters were mostly closed type which made up of bricks with cement slab roof. Collections were immediately transported to field laboratory used to be set up on every visit at the Karnal District, sorted according to species, identified based on standard keys (Sirivanakarn 1976, Reuben et al. 1994; "Pictorial key to common species of *Culex (Culex)* mosquitoes associated with Japanese encephalitis virus in India"—Chap. 3 of this book; Das et al. 1990) and female abdominal condition (unfed, freshly fed, semi gravid, and gravid) was recorded.

6.3.1 Prevalence of Cx. tritaeniorhynchus, the Principal JE Vector Species in the Karnal District

On 4 October 2002, in order to have preliminary information on adult mosquito prevalence in Karnal District, adult mosquito collection (outdoor, indoor, and dusk collection) was undertaken in village Tikri Kalan. Table 6.1 summarizes the results

Table 6.1 Results of adult mosquito collection from outdoor vegetation and cattle shed/human dwellings from Karnal District, Haryana (October 2002)

| SPECIES | Mosquitoes collected from outdoor vegetation | | | | | Mosquitoes collected from cattle shed/human dwellings | | | | | | |
| | | | | | | Morning collection | | | | | Evening collection | |
	Male (M)	Female (F)	Total	M: F ratio	Catch/10 hop cages	Male (M)	Female (F)	Total	M: F ratio	Catch/Man/Hour	Female	Catch/Man/Hour
Cx tritaeniorhynchus	207	424 (99.06)	631	1:2.05	45.6	2	10 (12.98)	12	1:5	20.0	180 (34.61)	90.0
Cx. bitaeniorhynchus	–	1 (0.23)	1	0:1	0.02	–	–	–	–	–	1 (0.19)	0.5
Cx. gelidus	–	2 (0.47)	2	0:2	0.04	–	–	–	–	–	–	–
Cx. infula	–	1 (0.23)	1	0:1	0.02	–	–	–	–	–	–	–
Cx. perplexus	–	–	–	–	–	1	5 (6.49)	6	1:5	10.0	214 (41.15)	107.0
Cx. quinquefasciatus	–	–	–	–	–	24	32 (41.56)	56	1:1.3	64.0	124 (23.84)	62.0
Anopheles subpictus	–	–	–	–	–	–	29 (37.66)	29	0:29	58.0	–	–
An. peditaeniatus	–	–	–	–	–	–	–	–	–	–	1 (0.19)	0.5
Mansonia uniformis	–	–	–	–	–	–	1 (1.3)	1	0:1	2.0	–	–
Total	207	428	635			26	77	104			520	

of adult mosquito survey in village Tikri Kalan during October 2002. A total of 428 and 77 female mosquitoes were collected in outdoor collection, resting in grassy ground vegetation of Jowar field and in indoor collection from cattle sheds/human dwellings, respectively. Dusk collection yielded 520 female mosquitoes belonging to four Culicine and one Anopheline species. Outdoor collections comprised of four culicine mosquitoes: *Cx. tritaeniorhynchus, Cx. bitaeniorhynchus, Cx. gelidus* and *Cx. infula.* Indoor (morning) collections revealed five species: two Anopheline (*An. subpictus, An. peditaeniatus),* two culicine species *(Cx. perplexus, Cx. quinquefqs-ciatus*), and one *Mansonia* species *(Mn. uniformis).* Of these *Cx. tritaeniorhynchus* was the most abundant JE vector species in outdoor collection comprising of 99.06 % of total mosquito collection while in indoors this species accounted for only 12.98 and 34.61 % of total mosquito collection during morning and evening hours, respectively. Predominance of *Cx. perplexus* (41.14 %), a nonvector species in dusk collection in the area requires careful attention as the species may be easily mistaken as *Cx. tritaeniorhynchus.* The study indicated *Cx. tritaeniorhynchus* was the most abundant JE vector species in the Karnal District, a typical feature reported in several JE affected areas in India (Soman et al. 1976; Mishra et al. 1984; Das et al. 2004; Kanojia et al. 2003; Kanojia 2007).

6.3.2 Seasonal Abundance of Cx. tritaeniorhynchus in Study Sites in Karnal District

Study of the seasonal abundance of vector populations is absolutely essential to develop suitable vector control strategies (Murty et al. 2002). Figure 6.4 shows correlation of JE vector density with climatic factors in Karnal District. The data revealed that seasonal abundance of *Cx. tritaeniorhynchus,* principal JE vector species follow closely rainfall pattern in the district and the peak of vector population of 103 per ten hop cages (PTHC) occurred during peak monsoon month (July). Figure 6.5 shows monthly averages for *Cx. tritaeniorhynchus* taken in outdoor and indoor collections at the study villages of the district during October 2002–September 2003. Abundance of the species could be monitored round the year under outdoor situation among land vegetation at all the study sites of the district while in indoor situation they could be collected in adequate numbers only during 4 months (July–October) September with peak observed in August [40.0 PMH]. However, 1.6X to 29.5X more of *Cx. tritaeniorhynchus* mosquitoes were collected in outdoors than that of indoor collection in Karnal District. More or less similar type of behavior was observed in affected villages of Saharanpur District (Fig. 7.11, Chap. 7).

Female specimens of *Cx. tritaeniorhynchus* in outdoor collection during April–October were in all stages of gonotrophic cycle, viz. unfed, fed, and gravid. From mid November onward, the night temperature in the area started falling and remained below 10 °C till end of February. *Cx. tritaeniorhynchus* females

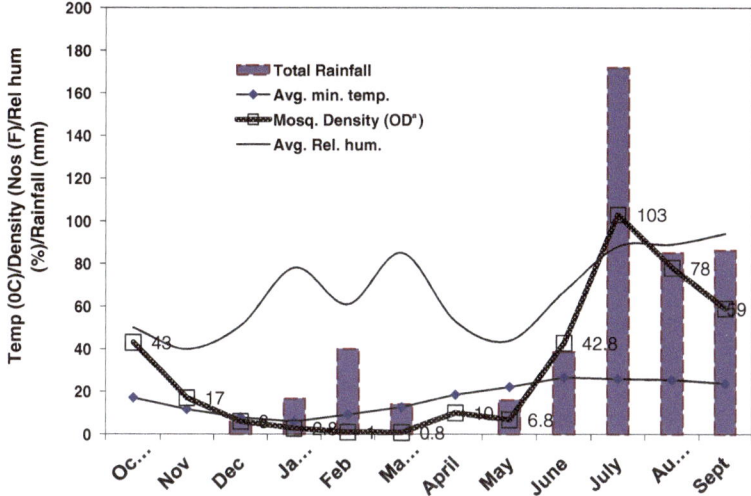

Fig. 6.4 Correlation of mosquito density (*Cx. tritaeniorhynchus*) in outdoor collection with climatic factors in district Karnal, Haryana. OD, Outdoor land vegetation, [a]Hop cage method

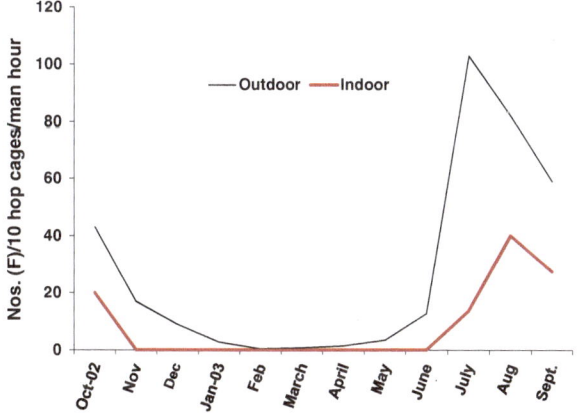

Fig. 6.5 Seasonal abundance of *Cx tritaeniorhynchus* in outdoor (hop cage) and indoor (hand catch) collection in study areas of Karnal district

suspended all its physiological activities during later part of November and remained quiescent in their outdoor resting locations in mustard and berseem till February. Therefore, only unfed population was available in the area that too in extremely low density (Fig. 6.6). The male population of *Cx. tritaeniorhynchus* disappeared during December–February from the area. Similar trend in vector bionomics was found in other parts of Northern India including Delhi, Haryana, Uttar Pradesh, and Uttaranchal where extreme cold climatic condition exists during winter months.

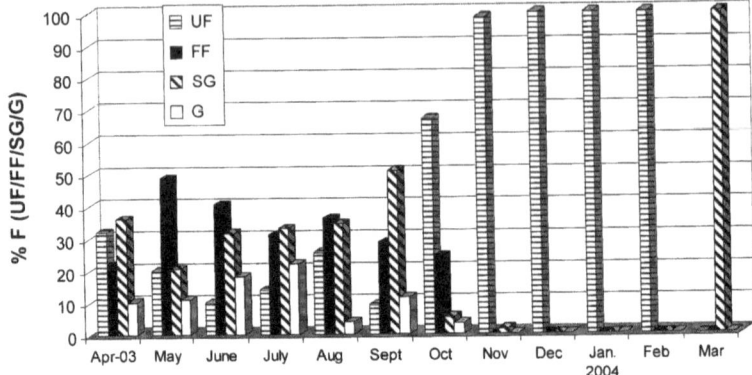

Fig. 6.6   Percentage of females (Unfed, Fed and Gravid) *Cx. tritaeniorhynchus* mosquitoes collected from outdoor vegetation of Karnal district (2003–2004)

6.3.3 Resting Habit of Vector Species in Study Areas of the District

During the day *Cx. tritaeniorhynchus* mosquitoes (males and females) were found to rest in different types local vegetation in different season of the year. Of these, the most preferred resting habitats were: Jowar (June–October), marigold (November–December), and berseem (December–May).

6.3.4 Seasonal Abundance of Larval Stages of Cx. tritaeniorhynchus in Karnal District

Paddy fields, ponds, and ground pools were found to contain breeding of *Cx. tritaeniorhynchus* in the district. Like in other parts of Northern India, male population of *Cx. tritaeniorhynchus* totally disappears during winter months (December–February). Vector population slowly builds up from March onward with ponds and ground water pools providing the scanty breeding of the species. The farmers in the area start paddy cultivation in the month of June with irrigated ground water available through bore wells. Once monsoon rain sets in during the month of July, the species profusely breeds in paddy fields till end of September when fields are left dry for the crop to be harvested in October. Paddy fields remained water logged for nearly 3 months and these were the major breeding source for the vector species (*Cx. tritaeniorhynchus*) during monsoon months (July–September) with per dip larval density ranging from 48, 80 to 90 during July, August, and September, respectively (Fig. 6.7b). After paddy was harvested in October, ponds with aquatic vegetation along the sides were found to contain appreciably high density of the species. Thus, *Cx. tritaeniorhynchus* was found to breed in ponds throughout the year except for three winter months (December, January, and February) with maximum density of 49 larvae per dip in

Fig. 6.7   Seasonal abundance of larval stages of *Culex tritaeniorhynchus* in Karnal district, Haryana (2003). **a** Pond with aquatic vegetation on sides. **b** Paddy fields

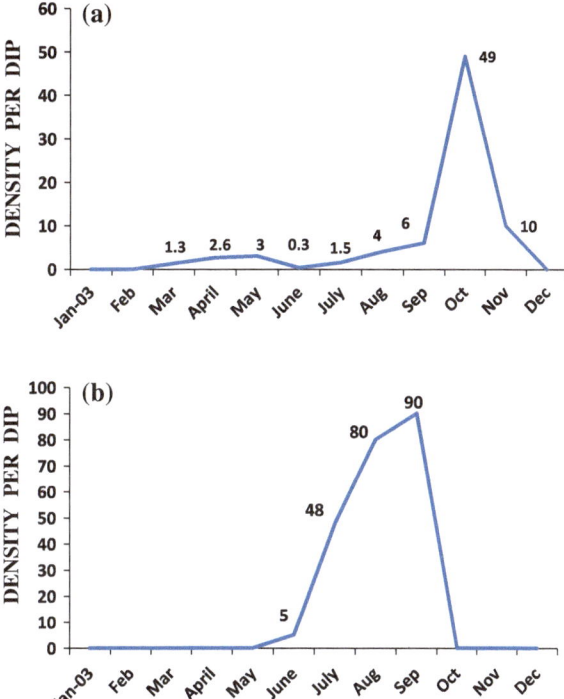

October and minimum being 0.3 in June (Fig. 6.7a). Negligible breeding of the species was recorded in ground pools in post monsoon months (October–November).

Similarly, *Cx. tritaeniorhynchus* breeding was recorded from both ground pool habitat and paddy fields in Mysore district and Karnataka state, India (Fakoorziba et al. 2006). In contrast, in another JE endemic district (Saharanpur) in Northern India paddy fields were found to provide only 5 % of total vector breeding in the area (Chap. 7). This type of differences in vector breeding preference is very important as far as planning of appropriate vector control strategies is concerned. There were few brick factories in the Karnal District but none were found with extensive rain water pools as were found in Saharanpur district.

6.3.5 First Time Detection of JE Virus in Vector Species from Haryana State

In an endemic area, surveillance of JE should be based on monitoring of JEV infection in natural population of vector mosquitoes (Tewari et al. 1999). Detection of JEV in vector mosquito is therefore an important aspect of the prevention of JE outbreaks as it can also detect circulation of JEV activity among zoonotic reservoirs and amplifiers (birds and pigs). However, detection of virus

Table 6.2 Wild caught adult mosquitoes from Karnal District (2002–2003) and tested for JE virus infection by ELISA[a]

Mosquito species	Locality	No. of pools tested/No. of adults/No. of positive
Culex tritaeniorhynchus	Karnal periurban area	**4/165/1**
	Kutail Kalan, PHC: Kutail	3/138/0
	Kambopura, Indri Block	2/100/0
	Karnal periurban area	2/81/0
	Kharkhali, PHC: Madhuban	4/168/0
Cx. perplexus	Tikri Kalan, Indri Block	2/135/0

[a]ELISA using detector antibody MAB 6B6C-1, cut off OD = 0.073

in mosquitoes depends upon the quality of preservation and storage of mosquitoes. Peiris et al. (1992) reported the use of antigen-capture enzyme immunoassay (EIA) for detection of JEV-antigen in field-collected mosquitoes stored at $-20\,°C$.

In India, due to the problems arising out of transportation, preservation, and storage of mosquitoes at sub-zero temperature; detection of virus is limited to a few laboratories located in south India. To overcome these problems Tewari et al. (1999) used unpreserved dry mosquitoes stored at room temperature for 30 days, and successfully detected JEV-antigen in *Cx. tritaeniorhynchus*. To make it more convenient, Das et al. (2005 report the detection of JEV-antigen in wild caught dry *Culex tritaeniorhynchus* stored at room temperature, without preservation, for a period of 20 months at NICD. In this method, wild caught adult mosquitoes in outdoor collections and dusk collections from study sites of Karnal District were separated in single-species pools (10–50 females per pool). These pools were transported in dry condition, in the absence cold chain facility to the centre for research in medical entomology (CRME), at Madurai, Tamil Nadu state, India, where they were processed for detection of JEV-antigen following the method of Gajanana et al. (1995).

A total of 17 pools (787 unfed female mosquitoes) representing two species were processed for detection of JEV infection by ELISA (Table 6.2). Of these 15 pools (652 mosquitoes) belonged to *Cx. tritaeniorhynchus* and two pools (135 mosquitoes) belonged to *Cx. perplexus*. A mosquito pool was considered positive for JEV antigen if its optical density (OD) was greater than or equal to mean + 4 standard deviation (SD) of the normal laboratory reared mosquito pools. Out of four pools of *Cx. tritaeniorhynchus* (165 mosquitoes) tested, one pool was found positive for JEV antigen (OD = 0.078). The cutoff OD was 0.073 (mean and SD of negative controls were 0.065855 and 0.00172240, respectively). OD value of positive control was 0.157 (Das et al. 2005). The minimum infection rate (MIR) of *Cx. tritaeniorhynchus* was 6.06 which was several times higher than that (0.84) reported earlier from Cuddalore district of Tamil Nadu, South India (Tewari et al. 1999). Other mosquito pools tested for virus infection were negative.

Review of JE cases from Karnal District revealed that first JE case was reported on 16 Oct 2002. The mosquitoes in the positive pool were collected just 10 days before the onset of the cases from cattle yard of National Dairy Research Institute

Fig. 6.8   Ideal day resting sites of vector species (*Culex tritaeniorhynchus*) among grasses (shown by *arrows*) in jowar crop

(NDRI) during evening hours in periurban locality of Karnal town and were stored at room temperature without preservation at NICD for 20 months. During 2002, one suspected JE case and one death were reported from Karnal town itself. This was the first record of JEV infection in vector mosquito from Haryana. It appears that timely detection of JE viral activity in vector mosquito during pre-transmission season in an endemic area followed by immediate suppression of infected vector population by appropriate vector control method might interrupt transmission of virus in human population and prevent child death due to Japanese encephalitis.

6.4   Spread of JE to New Areas

A complex interplay of ecological, environmental, climatic, and human behavioral factors has been implicated in the spread of JEV to new areas. These are: (i) As the regional economy grows, the promotion of more and more paddy fields in endemic areas may increase the risk of the disease both by attracting migratory birds (Cattle egrets, pond herons) that introduce the virus and by providing breeding grounds for the mosquitoes that transmit it, (ii) Rapid urbanization involving increased use of brick kiln industry right in the paddy growing areas may also contribute added risk of the disease by creating extensive low lying areas due to burrowing of earth and increasing JE vector breeding grounds and inviting more and more migratory birds (Chap. 7, Fig. 7.17), and (iii) Human cultural practice of growing millets (Jowar) in the adjoining areas of paddy fields during transmission season may also increase risk of the disease, by providing ideal day resting sites (Fig. 6.8) for both male and females of vector mosquitoes that transmit JEV.

6.5 JE Control Options

Control options for JE include: mosquito control, amplifying host (pig) control, and vaccination. However, each method has its own limitation and it is not practical to depend on any single method. Therefore, an integrated approach is suggested for prevention and control of Japanese encephalitis.

6.5.1 Mosquito Control

Conventional intervention methods like indoor residual spray with chemical insecticide, use bed nets, and treating extensive breeding source of the vector mosquito species (*Cx. tritaeniorhynchus*) in paddy fields with larvicides are not practical. As the vector mosquito mainly rest outdoors, indoor residual spray likely to have limited role. Bitting activity of *Cx. tritaeniorhynchus* mosquitoes was found to be confined from dusk period up to 22.00 h, use of bed nets at night is likely to be ineffective. Chemical larvicides have problems (e.g., mammalian toxicity, resistance, groundwater pollution, etc.), largely proved to be ineffective in reducing mosquito breeding in paddy fields. Bio-larvicides like *B.t.i* and entomopathogenic Nematodes are also ineffective, because the former cannot recycle in paddy fields and the later gets deactivated on being desiccated and exposed to ultraviolet light and agricultural pesticides.

6.5.1.1 Strengthening of Information Education and Communication Component at the District/PHC Level to Motivate People to Achieve the Following Objectives

These are: (i) Use of various physical and chemical methods for prevention of mosquito bite like mosquito repellents when outdoors, burning of dry leaves (preferably neem) during evening hours, etc., (ii) Use of full sleeve cloths, trousers, etc., to minimize mosquito bite, (iii) Mosquito proof piggeries should be built away from human habitations, (iv) Piggeries should be sprayed with an appropriate insecticide to reduce the vector density and to break the transmission cycle, (v) To eliminate adult resting sites (extensive low level ground vegetation in jowar) of important vector JE (*Cx. tritaeniorhynchus)*, and (vi) Promoting the use of appropriate biocontrol agents in paddy fields, ponds, etc., to reduce immature stages of the vector species.

6.5.1.2 Reduction of Vector Breeding Source

Eco-friendly methods to control vector breeding include: (i) Application of neem cake (a natural larvicide and fertilizer made from crushed neem nuts) in paddy fields, (ii) Treating paddy fields and other breeding sources of the vector species at the transplantation stage with a suitable bio-larvicide capable of recycling in the breeding sources of the vector, not unduly sensitive to ultra-violate radiation and vagaries of agricultural pesticides and can disperse in the environment via transovarian transmission. It may be mentioned that the active ingredient of the formulation *Chilodonella uncinata* BP 610, a new protozoan microbial control agent for mosquito larvae possesses many properties of a good biological control agent (Das 2008). Field trials are awaited for the formulation and once available this new biocontrol agent is best suited to control *Cx. tritaeniorhynchus* larvae breeding in paddy fields that hold water for months together in areas like Karnal District, and (iii) Placing larvivorous fish in paddy fields as these hold water for over 3 months.

6.5.1.3 Control of Adult Mosquitoes

Control of outdoor resting mosquitoes As the vector species was found to rest predominantly in the rank vegetation: in jowar for 4 months (June–October), with peak abundance in July and in mustard (November), attempts should be made to work out with the feasibility of killing day resting adults JE vectors mosquitoes in rank vegetation (jowar, mustard) by thermal fogging with technical malathion.

Control of indoor resting mosquitoes Based on the observation that: (i) A part (13 %) of total mosquito population that rested indoors (cattle shed including human dwelling) include vector population (*Cx. tritaeniorhynchus*) during early part of JE transmission season (October) in Karnal District (Table 6.1; Fig. 6.5), (ii) Considering the fact that under high density situation the zoophilic vector species is likely to become an indiscriminate feeder leading to increase in man–mosquito contact and likely transmission of JEV to human population, and (iii) JEV infection was detected from vector mosquitoes collected from NDRI Cattle yard, Karnal Town just 10 days before the onset of JE cases in the area. Therefore, it is advisable, as an *emergency control measure*, to undertake thermal fogging with technical malathion in human dwelling in known endemic villages during September, October, and first half of November to have a significant decrease in man–vector contact to prevent disease transmission.

6.5.1.4 Reduction in Man–Mosquito Contact

Since *Culex* mosquitoes preferably feed on cattle, which act as "dampers" in the natural cycle of JEV, cattle be used to divert infected mosquitoes from human. To reduce the risk of being bitten by infected mosquitoes include: (i) Minimizing the time spent outdoors in the evening, (ii) Wearing clothing that leaves minimal skin exposed, (iii) Liberal use of mosquito repellents, and (iv) Using long lasting insecticide impregnated curtain in doors and windows, (iv) burning of dry neem leaves during evening hours close to human dwelling from July to mid November to repel vector mosquitoes to minimize man–vector contact.

6.5.2 Control of Pigs

Vaccination of pigs has been used in other countries but has not been shown consistently reduce mosquito abundance and human infection; moreover, it is not cost effective intervention method. Moving pigs away from human habitation make sense, though pigs need to be moved at least 5 km away and the benefit has not been proved. Better option is: (i) Keeping pigs in mosquito proof piggeries, and (ii) It may be made compulsory to confine pigs during night hours in their pigsties or in the community pig sheds. Even, if legislation is required it may be passed, (iii) As on today no one knows how to manage and quarantine vagabond pigs of municipal areas in JE endemic districts, and (iv) Herds of pigs should not be allowed to move from one village to another during July–October to avoid pig-vector (epidemiologically dangerous *Cx. tritaeniorhynchus* population) contact in the district.

6.5.3 Vaccination

During 2007, Karnal District was covered under mass vaccination programme of JE with 90.01 % coverage (Operational Guidelines 2010). Children between age group 1 and 15 years were vaccinated with a single dose of Chinese vaccine SA14-14-2 strain.

The district was considered as an "endemic" area as per the classification of (Sabesan et al. 2008) where children (<15 years) alone are vulnerable to JEV infection; therefore, remaining children of the first batch (mass vaccination programme of JV carried out during 2007) who are still <15 years together with new comers from 2008 to 2012 are to be considered for future vaccination programme.

References

Das BP (2008) New microbial insecticide–a discovery by accident. Invent Intell 43:26–28

Das BP, Lal S, Saxena VK (2004) Outdoor resting preference of *Culex tritaeniorhynchus,* vector of Japanese encephalitis in Warangal and Karim Nagar district, Andhra Padesh. J Vector Borne Dis 41:32–36

Das BP, Rajagopal R, Akiyama J (1990) Pictorial key to the species of Indian Anopheline mosquitoes. J Pure Appl Zool 2:131–162

Das BP, Sharma SN, Kabilan L, Lal S et al (2005) First time detection of Japanese encephalitis virus antigen in dry and unpreserved *Culex tritaeniorhynchus* mosquitoes Giles, 1901, from Karnal District of Haryana state of India. J Commun Dis 37:131–133

Fakoorziba MR, Vijayan VA (2006) Seasonal abundance of larval stages of Culex species Mosquitoes (Diptera: Culicidae) in an endemic area of Japanese encephalitis in Mysore, India. Pak J Biol Sci 9:2468–2472

Gajanana A, Rajendran R, Thenmozhi V, Philip Samuel P et al (1995) Comparative evaluation of bioassay and ELISA for detection of Japanese encephalitis virus in field collected mosquitoes. Southeast Asian J Trop Med Public Health 26:91–97

Government of India (2010) Operational guide for Japanese encephalitis vaccination in india. Min. of Hith & F.W

Kar NJ, Saxena VK (1998) Some epidemiological characteristics of Japanese encephalitis in Haryana state of Northern India. J Commun Dis 30:129–131

Kanojia PC (2007) Ecological study on mosquito vectors of Japanese encephalitis virus in Bellary district, Karnataka. Indian J Med Res 126:152–157

Kanojia PC, Shetty PS, Geevarghese G (2003) A long-term study on vector abundance & seasonal prevalence in relation to the occurrence of Japanese encephalitis in Gorakhpur district, Uttar Pradesh. Indian J Med Res 117:104–110

Mishra AC, Jacob PG, Ramanujam S, Bhat HR et al (1984) Mosquito vectors of Japanese encephalitis epidemic (1983) in Mandya district (India). Indian J Med Res 80:377–389

Murty US, Satyakumar VR, Sriram K, Rao KM et al (2002) Seasonal prevalence of *Culex vishnui* subgroup, the major vector of Japanese encephalitis virus in an endemic district of Andhra Pradesh, India. J Am Mosq Control Assoc 18:290–293

Peiris JSM, Amerasinghe FP, Amerasinghe PH, Ratnayaka CB et al (1992) Japanese encephalitis in Sri Lanka—a study of an epidemic: vector incrimination, porcine infection and human disease. Trans R Soc Trop Med Hyg 86:307–313

Prasad SR, Kumar V, Marwaha RK, Batra KL et al (1993) An epidemic of encephalitis in Haryana: serological evidence of Japanese encephalitis in a few patients. Indian J Paediatr 30:905–910

Reuben R, Tewari SC, Hiriyan J, Akiyama J (1994) Illustrated key to genera of Culex (*Culex)* associated with Japanese encephalitis in Southeast Asia ((Diptera: Culicidae). Mosq Syst 26:75–96

Sabesan S, Konuganti HKJ, Perumal V (2008) Spatial delimination, forecasting and control of Japanese encephalitis: India—a case study. Open Parasitol J 2:59–63

Sharma SN, Panwar BS (1991) An epidemic of Japanese encephalitis in Haryana in the year 1990. J Commun Dis 23:204–205

Sharma RC, Saxena VK, Bharadwaj M, Sharma RS et al (1991) An outbreak of Japanese encephalitis in Haryana-1990. J Commun Dis 23:168–169

Sirivanakarn S (1976) Medical entomology studies III. A revision of the subgenus *Culex* in the oriental region (Diptera: Culicidae). Contrib Am Entomol Inst (Ann Arbor) 12(2):1–272

Soman RS, Kaul HN, Guru PY, Modi GB et al (1976) A report on the mosquitoes collected during an epidemic of encephalitis in Burdwan and Bankura districts, West Bengal. Indian J Med Res 64:808–813

Tewari SC, Thenmozhi V, Rajendran R, Appavoo NC et al (1999) Detection of Japanese encephalitis virus antigen in desiccated mosquitoes: an improved surveillance system. Trans R Soc Trop Med Hyg 93:525–526

Chapter 7
Ecology of Mosquito Vectors of Japanese Encephalitis Virus in Saharanpur District (JE Endemic Area), Uttar Pradesh State in Northern India

Abstract This chapter presents results of 1 year (2005–2006) longitudinal study on the ecology of vectors of Japanese encephalitis virus carried out in two epidemiologically distinct situations in Saharanpur District (JE endemic area), Uttar Pradesh state, in Northern India to develop appropriate vector control/ management strategy. The study revealed *Cx. tritaeniorhynchus* was the most abundant JE vector species that predominantly rested outdoors in vegetation and was found to change its resting habitat in accordance with the local agriculture practices. Paddy fields contributed approximately 5 % of total vector breeding, while rain water pools created by brick factories and hyacinth ponds together contributed 95 % of *Cx. tritaeniorhynchus* breeding in Saharanpur District. The differences between affected and unaffected villages of the district were a decreased abundance of *Cx. tritaeniorhynchus* mosquitoes during transmission period of the disease, presence of large fish pond and absence of brick factories in the unaffected villages. Saharanpur District has been reporting suspected cases of JE since 2002 and the outbreak of 2005 involved 221 cases of human encephalitis. In 2005, JE virus antigen was detected repeatedly from *Cx. tritaeniorhynchus* mosquitoes and vertical transmission of JE virus was established in 3 mosquito species (*Cx. tritaeniorhynchus*, *Cx. vishnui* and *Cx. fuscocephala*) in the district. When JE occurrence was analysed together with viral antigen detection (ELISA) in JE vectors, it was found that vertical transmission of JE virus occurred in two species, viz. *Cx. tritaeniorhynchus* and *Cx. vishnui*, 2 months prior to reporting of human encephalitis cases. This is an early warning signal for initiating integrated vector control measures to prevent JE outbreak.

7.1 Introduction

Saharanpur district forms the most northerly position of the Doab land (the flat alluvial tract between the holy rivers of the Ganges and the Yamuna in western and south western Uttar Pradesh and Uttarakhand state in India). The district is in a rectangular shape and lies between $29^0 58'$ and $30^0 21'$ North latitude and $77^0 9'$ and $78^0 14'$ East longitude. The north and the northeast portion of Saharanpur

B. P. Das, *Mosquito Vectors of Japanese Encephalitis Virus from Northern India*, SpringerBriefs in Animal Sciences, DOI: 10.1007/978-81-322-0861-7_7, © The Author(s) 2013

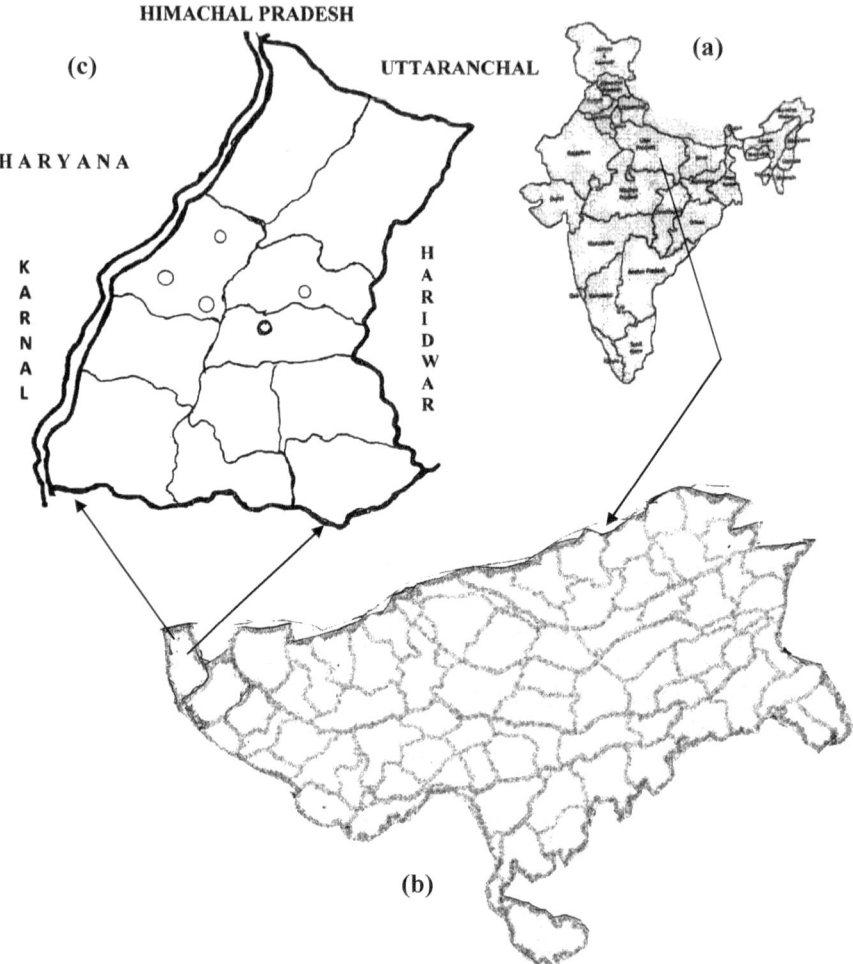

Fig. 7.1 a India. **b** Uttar Pradesh (District Map). **c** Saharanpur District showing administrative blocks and study sites (as "O")

District are surrounded by Shivalik hills and separate it from Dehradun District of Uttaranchal. River Yamuna forms its boundary in the west which separates it from Karnal and Yamunanagar districts of Haryana state. In the east lies Haridwar District which was a part of district Saharanpur before 1989 and in the south is district Muzaffarnagar of UP (Fig. 7.1). Saharanpur district can be divided into four parts: Shivalik Hill Tract, Bhabar Land, Khadar Land and Bangar Land, out of which Bhabar Land is adjacent to Shivalik hills. Yamuna is the important river of the district. Apart from this Solani, Hindon, Ratmau, Nagdev also run through the district and all these rivers submerge either in Yamuna or in the Ganges.

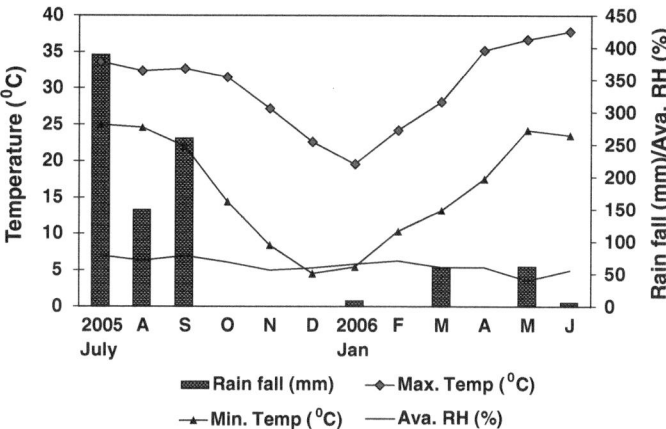

Fig. 7.2 Climatic condition of Saharanpur district

The district has a population of 3,028,104 as per district record maintained with DMO, Saharanpur, male to female ratio is 1,000:868. The district is divided into 11 administrative blocks with 37 primary health centres (PHCs), 297 subcentres, 1,307 villages and one district hospital (Seth Baldev Das Bajoria Hospital, Saharanpur) having 260 beds. In addition, there are 45 animal clinics and 37 hospitals for treatment of animals. Major towns in the district are Saharanpur, Behat, Deoband, Gangoh and Rampur. There are 5 city municipalities, 781 Gram Panchayats in the district. The climate of the district is dry subhumid to semiarid. Relative humidity ranges from 40 to 78 % and temperature ranging from −1 to 45 °C. The area receives on an average 935-mm rainfall, 86 % of which is received due to southwest monsoon during months of June to September. Broadly, there are three season—hot summer (April–June), hot and humid rainy season (July–October) and cool winter (November–March). May is the hottest month, maximum temperature on individual days sometimes reaches 45 °C. After November both day and night temperatures decrease rapidly till January which is the coldest month (Fig. 7.2). During cold season, the district is affected by cold wave and fog and minimum temperature occasionally goes down to −1 °C.

Common vector-borne diseases endemic to this district are malaria and Japanese encephalitis. The district is reporting seasonal (September–December) outbreaks of JE cases regularly since 2002 with high case fatality rate. In most of these outbreak investigations, adult mosquito survey indoors had shown negligible abundance of JE vector species in the area. As a result, there was much confusion about the reason for the outbreak and the disease transmission could not be explained on the basis of entomological evidence. During November 2004 outbreak in Saharanpur, for the first time adult mosquito survey outdoors using BPD hop cage method showed moderate presence of *Cx. tritaeniorhynchus*, principal vector of JE, as against a total absence of the species from indoor locations. Following the same sampling tool, *Cx. tritaeniorhynchus* accounted for 93 % of total mosquito collection from outdoor vegetation in contrast to poor abundance (2.9 %) of this species from indoor collection in viral encephalitis outbreak during July 2003 in Warrangal and Karimnagar district of Andhra Pradesh, (Das et al. 2004). Similar trend was seen in *Cx. tritaeniorhynchus* abundance in JE outbreak investigation during July 2004 in Gorakhpur District in Eastern UP (Gupta et al. 2005).

7.1.1 Mosquito Control Strategy Adopted by Saharanpur District

The data maintained at the District Malaria Office, Saharanpur revealed that fever rate of the district was 0.6, 0.15 and 0.15 % in 2003, 2004 and 2005, respectively. Since, Annual Parasite Rate (API) was >2 for many years, the district happened to be not under insecticidal spray operation under national vector borne disease control programme (NVBDCP). However, the District Malaria Office carried out thermal fogging once a case of encephalitis reported from a village.

7.1.2 Need for Ecological Study of JE Vectors in Saharanpur District

Regarding JE vectors, there are many gaps in the knowledge on different aspects of ecology of vectors of JE virus in India. As such indoor residual spray of insecticide is not very effective due to the exophilic habit of the vector mosquitoes (Reuben 1971a, b, c; Das et al. 2004). In formulating control strategies for vectors of Japanese encephalitis virus, information on the ecology of these mosquitoes are very much essential. In view of the above, 1-year longitudinal study was undertaken on the ecology of JE vectors from July 2005 to June 2006 in two groups of epidemiologically distinct areas in Saharanpur district: one affected with JE and the other that remained free from the disease till 2004 to develop appropriate JE vector surveillance tool to formulate situation specific vector control strategies. This chapter presents: (i) Epidemiological profile of Japanese encephalitis in the state/district, (ii) Mosquito fauna Saharanpur district, including notes on their medical importance, (iii) Comparative seasonal abundance of *Cx. tritaeniorhynchus* from affected and unaffected villages of Saharanpur district, (iv) Shifting pattern in the resting habits of *Cx. tritaeniorhynchus* from one crop to another in Saharanpur district, (v) Feeding behaviour of *Cx. tritaeniorhynchus* in the district, (vi) Brick kiln industry and its role in JE vector breeding potential in Saharanpur district, (vii) Detection of natural vertical transmission of JE virus in vector mosquitoes of Saharanpur district, (viii) Seasonal variations in minimum infection rates of JE virus in vector mosquitoes in relation to seasonal abundance of

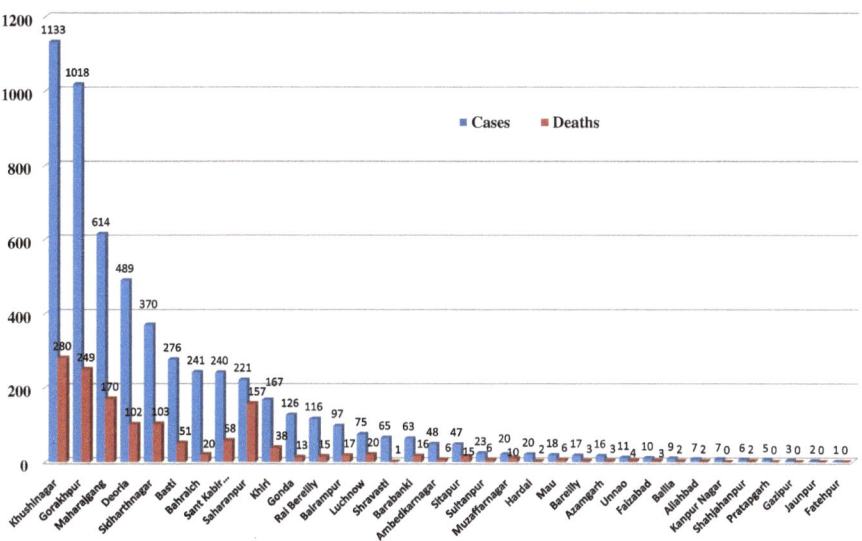

Fig. 7.3 District wise distribution of cumulative cases and deaths from JE in UP during 2005 (*Source* NICD)

Cx. tritaeniorhynchus in Saharanpur district, and (ix) JE vector surveillance and feasible situation specific vector control/management strategies to prevent outbreak of the disease.

7.2 Epidemiological Profile of Japanese Encephalitis in Uttar Pradesh State/Saharanpur District

7.2.1 Magnitude of JE Problem in Uttar Pradesh State

During the year 2005, a total of 5,581 suspected cases of Japanese encephalitis with 1,387 deaths were reported from UP state of India. The overall case fatality rate was 24.85 %. An analysis of the district profile (Fig. 7.3) shows that 12 districts have reported more than 100 cases in that year. These were: Kushinagar (1,133 cases and 280 deaths), Gorakhpur (1,018 cases and 249 deaths), Maharajganj (614 cases and 170 deaths), Deoria (489 cases and 102 deaths), Siddharthnagar (370 cases and 103 deaths), Basti (276 cases and 51 deaths), Bahraich (241 cases and 20 deaths), Sant Kabir Nagar (240 cases and 58 deaths), Saharanpur (221 cases and 157), Khiri (167 cases and 38 deaths), Gonda (126 cases and 13 deaths) and Rai Bareilly (116 cases and 15 deaths). Of these 12 districts, Saharanpur witnessed highest (71.04 %) case fatality rate (CFR) as against overall state case fatality rate of 24.85 % (Fig. 7.4). Saxena et al. (2009)

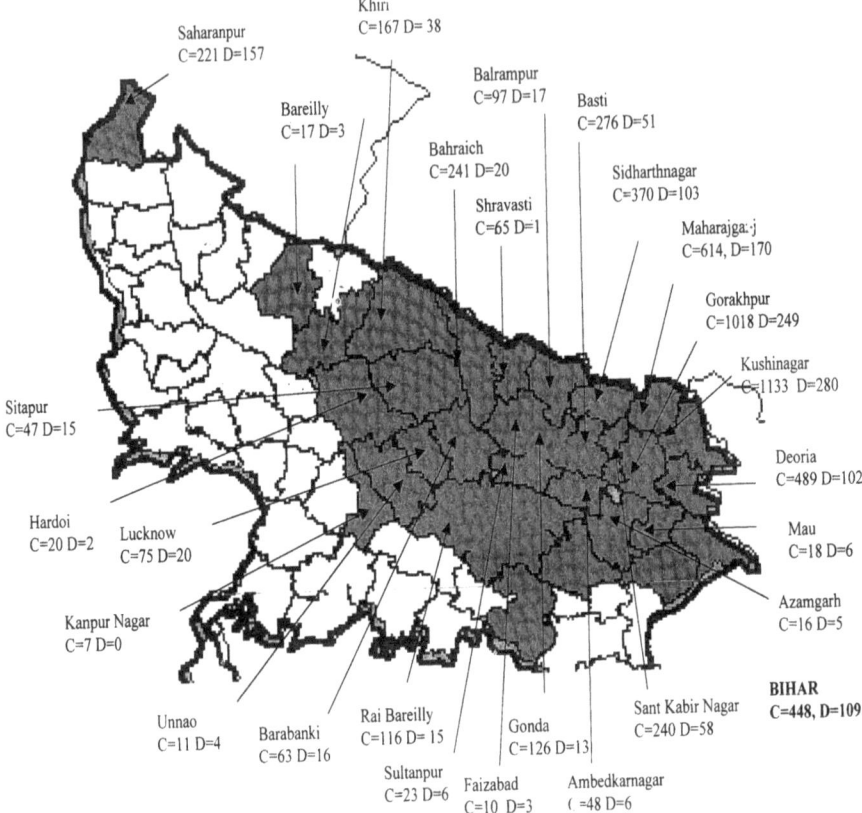

Fig. 7.4 Geographical distribution of suspected cases and deaths of Japanese encephalitis in UP during 2005 (*Source* NICD)

reconfirmed the circulation of JE in the massive outbreak of Japanese encephalitis that occurred in the year 2005 in northern states of India.

7.2.2 Japanese Encephalitis Problem in Saharanpur District

Saharanpur District is reporting suspected cases of JE regularly since the year 2002. The review of the paediatric ward records maintained at District Hospital, Saharanpur (Table 7.1) showed that the admission due to suspected JE cases began in the month of October during the initial 2 years (2002–2003). In 2004–2005, the admission of cases began in the month of September and it was also observed that there were an increased number of admissions in the month of October–November. The transmission season of JE in Saharanpur extends from September to December (Fig. 7.5).

Table 7.1 Month wise incidence of hospitalised cases[a] and deaths suspected for JE (2002–2005)

Years	Month wise cases and deaths										CFR (%)
	Jan–August		September		October		November		December		
	Cases	Death	Cases	Death	Cases	Death	Cases	Death	Cases	Death	
2002	0	0	0	0	18	16	22	24	0	0	100
2003	0	0	0	0	56	53	40	26	11	11	84.11
2004	0	0	8	5	80	59	78	48	1	1	67.66
2005	0	0	13	9	91	64	98	76	19	8	71.04

[a]Majority (92.3 %) of these cases pertains to age group 1–10 years admitted in Paediatric Ward of Seth Baldev Das Bajoria Hospital, Saharanpur District. Only 7.7 % of cases belong to age group >10 years

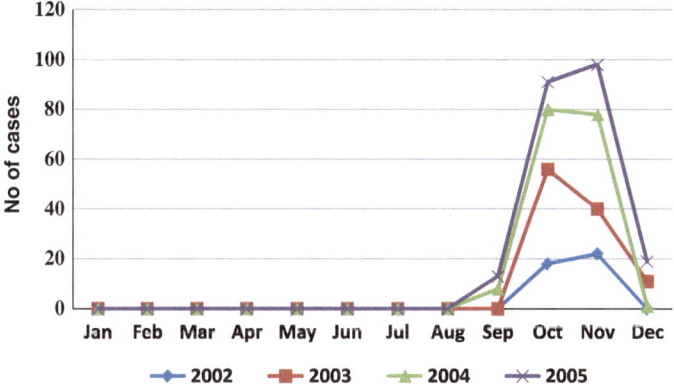

Fig. 7.5 Line diagram showing incidence of suspected JE cases at monthly interval admitted to Paediatrics ward of Saharanpur District hospital (2002–2005)

Of the 1,307 villages of Saharanpur district, 176 villages (population of 7,67,722) reported suspected cases of JE in 2005. Epidemiological data of suspected JE cases during the last 4 years (2002–2005) was collected from the District Malaria Office, Saharanpur district. During 2005, first suspected case of JE was admitted at Saharanpur district hospital on 15 September 2005. Thereafter, one or two patients were kept on being admitted on each day with hardly 1–2 days gap in a week and the trend continued till December with peak in November. Maximum number >(85 %) of cases occurred during October and November. A total of 221 cases including 157 deaths with a case fatality rate of 71.04 % were reported in 2005. Sudden onset of high fever and rapid death were the characteristics of the disease in many cases. However, some had moderate fever for 4–6 days, with intermittent altered sensorium, vomiting, seizures, semiconsciousness and coma before death. The outbreak of 2005 involved the neighbouring Muzaffarnagar district, with 21 cases. Distribution of reported cases indicated that cases were scattered all over Saharanpur district as observed elsewhere (Mani et al. 1991).

Table 7.2 Age and sex distribution of suspected JE cases in Saharanpur district

Age group (Year)	Number of JE cases			Percentage (%)
	Male	Female	Total	
<1 year	–	–	–	–
1–<3	20	11	31	14.02
3–<5	32	47	79	35.75
5–< 10	40	53	93	42.08
10–15	12	6	18	8.14
Total	104	114	221	–
%	47.05	51.58	–	99.99

Source District Health Office, Saharanpur

The affected villages in a Block did not show any clustering, with on an average one case per village.

Out of 221 cases 110 (49.77 %) cases were under 5 years of age, while 93 (42.08 %) cases were between 5 and 9½ years of age and 18 (8.14 %) cases were between 10 and 15 years of age. There were no JE cases found in infants. Male children below 3 years and above 10 years were predominantly affected by JE in comparison to female (Table 7.2). The attack rate of suspected encephalitis in Saharanpur district was found to be 0.2–0.3 per 10,000 populations (Fig. 7.6a). Cases of JE were recorded from all the blocks including municipal areas of Saharanpur district. The affected villages in a block did not show any clustering. One or two cases were reported from an affected village. However, maximum number of cases was reported from Sarsawa and Gangoh block that are adjacent to the river Yamuna (Fig. 7.6b).

Serum and CSF samples collected from patients in 2005 were sent to NICD in batches by DMO, Saharanpur for detection of JE virus. Out of 115 serum and 108 CSF samples tested at NICD, three serum samples and four CSF samples were tested positive for JE virus. These records are sufficient to confirm that even though, all the cases and deaths diagnosed at Saharanpur district may not be due to JE but there was no doubt that occurrence of local transmission of JE virus in the area. This was further substantiated by the entomological evidence collected during the study period and included in this chapter.

7.3 Mosquito Fauna of Saharanpur District Including Notes on their Medical Importance

7.3.1 Topography of Saharanpur District

Saharanpur is very fertile agricultural belt famous for plentiful yield in grains and fruits. Although Dehradun is more famous for basmati rice, a lot of it is grown in the Saharanpur area. The main crops in the area are sugarcane

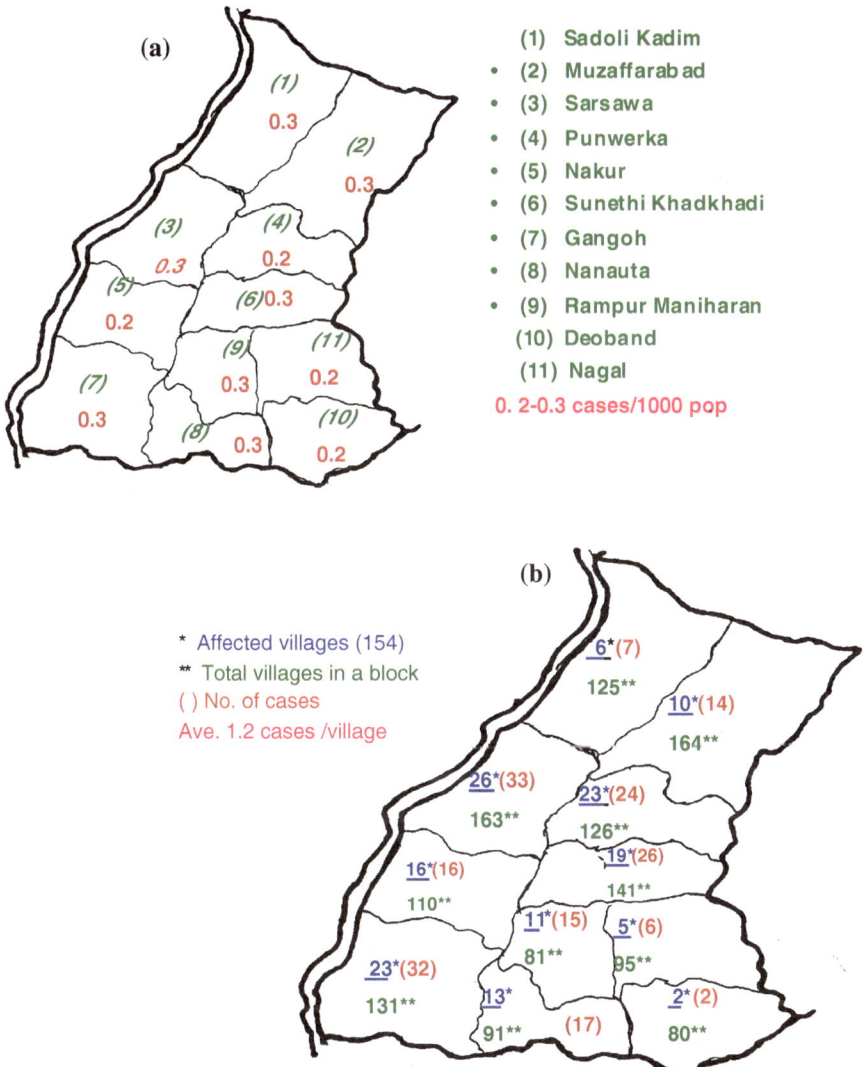

Fig. 7.6 a Attack rates (No. per 10,000 population) of suspected JE cases in Saharanpur District (2005) [Name of block is indicated in bracket]. **b** Block wise distribution of suspected JE cases in Saharanpur district (2005) [No. of affected villages indicated (*), total villages in a block (**), No. of cases ()]

(*Saccharum officinarum* Linn.), paddy (*Oryza sativa* Linn.), wheat (*Triticum aestivum* Linn.) and jowar (*Sorghum bicolor* Linn.) while minor crops include mustard *Brassica juncea* (Linn.) and berseem (*Trifolium alexandrinum* Linn.). Sugar cane is the annual crop in cultivated land area of 39.9 % of the district. Broadly, paddy and jowar (millet) are grown in close proximity during monsoon

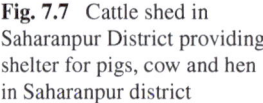

Fig. 7.7 Cattle shed in
Saharanpur District providing
shelter for pigs, cow and hen
in Saharanpur district

and post-monsoon months (June–October) with land area of 16.3 % and 10.7 %,
respectively. Wheat and berseem (fodder plant) are grown from December to
following April while mustard crop is grown for 2 months (November and
December). The millet and mustard crops support extensive low, ground grassy
vegetation where mosquitoes prefer to rest during the day. The fodder plant is
low level ground vegetation and provides resting habitats to *Cx. tritaeniorhynchus* and *Cx. quinquefasciatus* mosquitoes particularly during winter months. In
addition, there are marshes with profuse growth of hyacinth plant (*Eichhornia
crassipes* Solms.) that support day resting habitat for *Cx. tritaeniorhynchus* mosquitoes as in many places of India including Delhi and Saharanpur.

Many villages (64 %) of the district have large permanent ponds in close
proximity to human habitation and there is practice of culturing edible fish in
large ponds. In addition, there are extensive marshes with profuse growth of
hyacinth plant and elephant grass. In some villages, there are temporary depres-
sions created due to digging of earth by the brick kiln industry which often
remained filled with water and aquatic vegetation during monsoon and post-
monsoon months. The adjacent fields of jowar, mustard and berseem provided
ideal day resting sites for the adult mosquitoes. The overall housing condition
has been found to be quite unsatisfactory with large number of kutcha houses
and only 37.44 % of households own pucca houses as per estimates of Govt
of India report. Invariably, in each house the living room is in close proximity
of cattle shed/poultry house/pig shelter. Cattle sheds are predominantly made of
brick wall and semiconcrete floor with asbestos roof. Sometimes, a cattle shed
was found to accommodate cow, pig and hen as well (Fig. 7.7). Occasionally,

Fig. 7.8 Cow dung cakes
stored in living room in
Saharanpur district

living rooms also found to store cow dung cakes that provide ideal resting sites
for indoor resting mosquitoes including *Culex tritaeniorhynchus*, primary vector
of JE in India, thereby increasing the risk of man–vector contact during the transmission season (Fig. 7.8).

7.3.2 Study Sites

Sample mosquito collections, both adult and immature, were made once in a
month between July 2005 to June 2006 in two groups of villages, viz. one that
reported at least one suspected case of JE since 2002 and the other that remained
free from the disease till the commencement of the study. The former group of villages were designated as "affected" and the later as "unaffected". Three affected
villages, viz. Chilkana, Budhakhera Ahir and Manak Mau belonged to Sarsawa,
Punwarka and Sunethi Kharkhari blocks, respectively (Fig. 7.1). Two unaffected
villages, viz. Halalpur and Pilakhni belong to Sarsawa block. During an earlier study on Malariogenic stratification of district Saharanpur carried out by the
author in 1991, these three affected blocks were found to have low to moderate
malariogenic potential, moderate to high malaria incidence and epidemic potential. The study villages were 10–40 km apart from each other and were found to
have high (80–90 %) irrigation potential.

7.3.3 Features of Affected and Unaffected Study Sites

The ecological surveys undertaken in affected and unaffected villages showed that they were almost similar in respect to agricultural crops. However, they differed in other respects, viz. The affected study villages were bigger, economically better off as indicated by the type of houses with modern day facilities and usually had a few brick kiln industry operating in dry season. Per capita cattle population of these villages was lower (0.16) and pig population was negligible. In contrast, unaffected study villages were comparatively smaller, majority of the population were from lower socioeconomic group. These villages had one large permanent pond used for edible fish culture but had no brick kiln industry. Per capita cattle population of these villages was higher (0.41) with comparatively more pig population. As per district record there are 1,882 pig shelters in Saharanpur district.

7.3.4 Adult Mosquito Survey

Adult mosquitoes were collected at monthly interval from five villages, viz. Chilkana, Manak Mau, Budhakhera Ahir, Halalpur and Pilakhni of Saharanpur district of UP State of Northern India and categorised as: Outdoor resting collection (hop cage) from 09:00 to 13:00 h and Indoor resting collections, hand catch (HC) from 06:00 to 0.8:00 h, Total catch (TC) from 10:00 to 12:00 h and dusk collection (DC) from 18:00 to 20:00 h. Collections were immediately transported to field laboratory set-up on every visit at the Postal Training College, Saharanpur; sorted according to species, identified based on standard keys (Sirivanakarn 1976; Reuben et al. 1994; "Pictorial key to common species of Culex (Culex) mosquitoes associated with Japanese encephalitis virus in India"—Chap. 3 of this book; Das et al. 1990) and female abdominal condition (unfed, freshly fed, semigravid and gravid) was recorded.

A total of 20 species of mosquitoes, belonging to 7 genera: *Aedes* (two species), *Anopheles* (5), *Armigeres* (1), *Culex* (8), *Mansonia* (2), *Neomelaniconion* (1) and *Verralina* (1) were collected from Saharanpur district (Table 7.3). *Culex* mosquitoes were the most dominant representing 84.8 % of the total collections followed by *Neomelaniconion* (7.3 %), *Anopheles* (6.3 %), *Aedes* (0.8 %), *Armigeres* (0.4 %), *Verralina* (0.2 %) and *Mansonia* (0.1 %) (Fig. 7.9). Table 7.3 summarises the collections of all species made by different collection methods in all the study villages during 2005–2006. Culicine mosquitoes collected were mainly exophilic. *Cx. tritaeniorhynchus* predominated in outdoor collections in all the study sites comprising 66.6 % of the total mosquito collected in Chilkana, followed by 65.2 % (Halalpur), 59.7 % (Pilakhni), 59.3 % (Budhakhera Ahir) and 54.4 % (Manak Mau). Its prevalence in indoor situation was (6.0–29.4) % during 18:00–20:00 h and <4.0 % during 06:00–08:00 h and 10:00–12:00 h, respectively.

The list of collected mosquitoes from Saharanpur includes seven medically important species in the area, viz. *An. culicifacies, An. stephensi, Cx. fuscocephala, Cx. gelidus, Cx. pseudovishnui, Cx. tritaeniorhynchus* and *Cx. vishnui*. Of these, first two

Table 7.3 Mosquito species collected from Saharanpur district (2005–2006)

Villages	Species	Indoor						Outdoor	
		(HC) 06–08 h		(TC) 10–12 h		(DC) 18–20 h		BPD hop cage (09–13 h)	
		Nos.[a]	%	Nos.[a]	%	Nos.[a]	%	Nos.[a]	%
Chilkana	Ae. jamesi (Edwards)	6	2.23	0	0	40	4.44	0	0
	An. culicifacies Giles	4	1.48	16	2.33	2	0.22	0	0
	An. peditaeniatus(Leicester)	0	0	0	0	43	4.77	0	0
	An. stephensi Liston	1	0.37	0	0	0	0	0	0
	An. subpictus Grassi	0	0	10	1.46	5	0.55	0	0
	Armigeres subalbatus(Coquillett)	0	0	1	0.14	9	0.99	1	0.62
	Culex bitaeniorhynchus Giles	1	0.37	1	0.14	8	0.88	0	0
	Cx. fuscocephala Theobald	0	0	0	0	2	0.22	22	13.83
	Cx. gelidus Theobald	0	0	0	0	0	0	7	4.40
	Cx. perplexus Leicester	0	0	0	0	56	6.21	5	3.14
	Cx. pseudovishnui Colless	0	0	0	0	17	1.88	1	0.62
	Cx. quinquefasciatus Say	254	94.42	614	89.63	464	51.49	14	8.80
	Cx. tritaeniorhynchus Giles	0	00	23	3.35	124	13.76	106	66.66
	Cx. vishnui Theobald	0	0	3	0.43	5	0.55	3	1.88
	Neomelaniconion lineatopennis(Ludlow)	1	0.37	17	2.48	126	13.98	0	0
	Verralina indica(Edwards)	2	0.74	0	0	0	0	0	0
	Total	269	100	685	100	901	100	159	100
Manak	Ae. albopictus (Skuse)[b]	2	0.55	0	0	0	0	0	0
Mau	Ae. jamesi	0	0	0	0	0	0	2	0.8
	An. annularis Van der Walp	0	0	0	0	0	0	2	0.8
	An. peditaeniatus	0	0	3	0.28	0	0	10	4.0
	An. stephensi	0	0	1	0.09	0	0	0	0

(continued)

Table 7.3 (continued)

Villages	Species	Indoor						Outdoor	
		(HC) 06–08 h		(TC) 10–12 h		(DC) 18–20 h		BPD hop cage (09–13 h)	
		Nos.[a]	%	Nos.[a]	%	Nos.[a]	%	Nos.[a]	%
	An. subpictus	30	8.33	80	7.52	8	0.98	0	0
	Ar. subalbatus	2	0.55	0	0	6	0.73	1	0.4
	Cx. bitaeniorhynchus	0	0	0	0	6	0.73	2	0.8
	Cx. fuscocephala	0	0	0	0	0	0	5	2.4
	Cx. gelidus	0	0	0	0	12	1.47	8	3.2
	Cx. perplexus	23	6.39	26	2.44	0	0	30	12.0
	Cx. pseudovishnui	0	0	0	0	11	1.34	2	0.8
	Cx. quinquefasciatus	300	83.33	947	89.08	569	69.64	45	18.0
	Cx. tritaeniorhynchus	2	0.55	6	0.56	190	23.25	136	54.4
	Cx. vishnui	1	0.27	0	0	6	0.73	7	2.8
	N. lineatopennis	0	0	0	0	1	0.12	0	0
	V. indica	0	0	0	0	8	0.98	0	0
	Total	360	100	1063	100	817	100	250	100
Budha-khera	Ae. jamesi	0	0	0	0	27	3.67	0	0
Ahir	An. annularis	0	0	2	0.27	4	0.54	0	0
	An. culicifacies	1	0.86	3	0.41	0	0	0	0
	An. peditaeniatus	0	0	0	0	22	2.99	2	2.19
	An. stephensi	1	0.86	0	0	0	0	0	0
	An. subpictus	41	35.34	200	27.51	8	1.08	0	0
	Ar. subalbatus	0	0	0	0	12	1.63	0	0
	Cx. bitaeniorhynchus	0	0	0	0	6	0.81	0	0
	Cx. perplexus	1	0.86	56	7.70	0	0	8	8.79
	Cx. pseudovishnui	0	0	2	0.27	62	8.43	9	9.89

(continued)

Table 7.3 (continued)

Villages	Species	Indoor						Outdoor	
		(HC) 06–08 h		(TC) 10–12 h		(DC) 18–20 h		BPD hop cage (09–13 h)	
		Nos.[a]	%	Nos.[a]	%	Nos.[a]	%	Nos.[a]	%
	Cx. quinquefasciatus	62	53.44	456	62.72	396	53.87	16	17.58
	Cx. tritaeniorhynchus	4	3.45	7	0.96	87	11.83	54	59.34
	Cx. vishnui	6	5.17	1	0.13	15	2.04	2	2.19
	Mansonia annulifera (Theobald)	0	0	0	0	3	0.4	0	0
	Mn. uniformis (Theobald)	0	0	0	0	4	0.54	0	0
	N. lineatopennis	0	0	0	0	89	12.10	0	0
	Total	116	100	727	100	735	100	91	100
Halalpur	An. annularis	1	0.48	0	0	4	0.36	0	0
	An. culicifacies	0	0	1	0.13	0	0	0	0
	An. peditaeniatus	0	0	2	0.26	0	0	1	0.39
	An. stephensi	0	0	0	0	2	0.18	0	0
	An. subpictus	42	20.09	16	2.09	3	0.27	0	0
	Ar. subulbatus	0	0	2	0.26	1	0.09	0	0
	Cx. bitaeniorhynchus	0	0	0	0	6	0.55	2	0.79
	Cx. gelidus	0	0	0	0	15	1.37	15	5.92
	Cx. perplexus	7	3.34	4	0.52	33	3.02	27	10.67
	Cx. pseudovishnui	4	1.91	6	0.78	30	2.76	0	0
	Cx. quinquefasciatus	150	71.77	675	88.46	668	61.28	41	16.20
	Cx. tritaeniorhynchus	1	0.48	5	0.65	321	29.45	165	65.21
	Cx. vishnui	0	0	0	0	6	0.55	1	0.39
	N. lineatopennis	4	1.91	52	6.81	1	0.09	1	0.39
	Total	209	100	763	100	1090	100	253	100

(continued)

Table 7.3 (continued)

Villages	Species	Indoor						Outdoor	
		(HC) 06–08 h		(TC) 10–12 h		(DC) 18–20 h		BPD hop cage (09–13 h)	
		Nos.[a]	%	Nos.[a]	%	Nos.[a]	%	Nos.[a]	%
Pilakhni	Ae. jamesi	0	0	0	0	7	0.75	0	0
	An. annularis	0	0	0	0	2	0.21	0	0
	An. culicifacies	2	1.20	39	5.72	0	0	0	0
	An. peditaeniatus	0	0	0	0	9	0.97	2	1.43
	An. subpictus	6	3.61	40	5.87	1	0.10	0	0
	Ar. subalbatus	0	0	0	0	3	0.32	6	4.31
	Cx. bitaeniorhynchus	1	0.60	1	0.14	1	0.10	1	0.71
	Cx. gelidus	0	0	1	0.14	10	1.08	0	0
	Cx. perplexus	0	0	6	0.88	13	1.40	1	0.71
	Cx. pseudovishnui	0	0	0	0	26	2.81	2	1.43
	Cx. quinquefasciatus	156	93.97	589	86.49	319	34.56	26	18.70
	Cx. tritaeniorhynchus	1	0.60	5	0.73	56	6.06	83	59.71
	Cx. vishnui	0	0	0	0	2	0.21	12	8.63
	Mn. annulifera	0	0	0	0	3	0.32	0	0
	N. lineatopennis	0	0	0	0	458	49.62	6	4.31
	V. indica	0	0	0	0	13	1.40	0	0
	Total	166	100	681	100	923	100	139	100

[a]Female

[b]Collected from a cooler

Fig. 7.9 Genus-wise
abundance of mosquitoes of
Saharanpur district

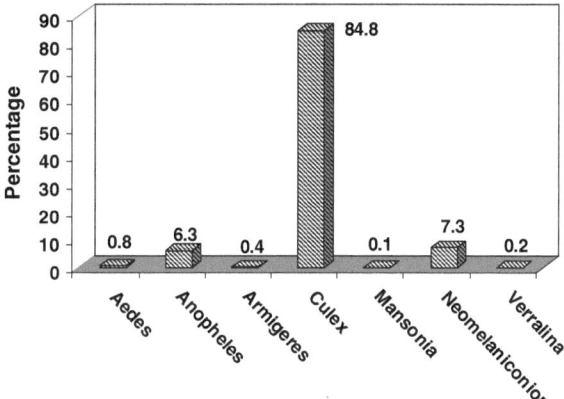

were encountered in indoor resting collections (HC, TC and DC), whereas rest five species were mainly found in outdoor resting collections employing BPD hop cage method (Das 2000, 2009). DC contributed 43.2 % of total collections, while TC, HC and hop cage method contributed 37.9, 10.8 and 8.04 %, respectively (Table 7.4).

A total of 831 mosquitoes in 4 h and 16 min were collected using BPD hop cage method from five villages under outdoor situation using 1,280 hopping attempts at 5 hopping attempts/minute/vegetation during 09:00–13:00 h. In contrast, 1,118 mosquitoes were collected in 10 h during 06:00–08:00 h from human dwellings using a mouth aspirator and torch light at 10 min/house, 3,919 female mosquitoes were collected in 40 h during from 10:00 to 12:00 h from HD using pyrethrum space spray at 30 min/per room/village and 4,466 female mosquitoes were collected in 45 h during 18:00–20:00 h around cattle shed and human dwelling using a mouth aspirator and torch light at 30–35 min/village. Of the total mosquitoes collected in outdoors comprising of 13 species, *Cx. tritaeniorhynchus* was the dominant species (61.6 %) followed by *Cx. quinquefasciatus* (17.7 %), *Cx. perplexus* (8.5 %), *Cx. gelidus* (3.6 %), *Cx. fuscocephala* (0.6 %), *Cx. vishnui* (2.8 %) and *An. peditaeniatus* (1.9 %) (Table 7.4). Jowar, berseem and mustard provided outdoor resting shelters for these mosquitoes.

Of the total 1,118 females collected from indoors consisting of 14 species, *Cx. quinquefasciatus* was the most dominant species (82.5 %) followed by *An. subpictus* (10.6 %), *Cx. perplexus* (2.8 %), *Cx. tritaeniorhynchus* (0.7 %), *Cx. vishnui* (0.6 %) and *An. culicifacies* (0.6 %). Comparatively more (3,919) mosquitoes comprising 13 species were collected from indoors by TC method in about 30 h. *Cx. quinquefasciatus* was the dominant (83.7 %) followed by *An. subpictus* (8.8 %), *Cx. perplexus* (2.3 %), *Neomelaniconion lineatopennis* (1.8 %), *An. culicifacies* (1.5 %) and *Cx. tritaeniorhynchus* (1.2 %). Dusk collection yielded highest number (19) of species with 4,466 female mosquitoes spending 45 h in collecting them. *Cx. quinquefasciatus* was the dominant (54 %) followed by *Cx. tritaeniorhynchus* (17.3 %), *Neomelaniconion lineatopennis* (15 %), *Cx. pseudovishnui* (3.2 %), *Cx. perplexus* (2.3 %), *An. peditaeniatus* (1.6 %) and *Ae. jamesi* (1.6 %).

Table 7.4 Species composition in female mosquitoes collected by various methods in Saharanpur district (2005–2006)

| | Outdoor resting | | Indoor resting | | | | | |
| | BPD Hop Cage (9–13 h)[a] | | Hand catch (6–8 h)[b] | | Total catch (10-h)[c] | | Dusk collection (18–20 h)[d] | |
	Nos. (%)	No./h	Nos. (%)	No./h	Nos. (%)	No./h	Nos. (%)	No./h
Aedes jamesi (Edwards)	2 (0.24)	0.48	6 (0.54)	0.6	–	–	74 (1.66)	1.64
Anopheles annularis Van der Walp	2 (0.24)	0.48	1 (0.09)	0.1	–	–	10 (0.22)	0.22
An. culicifacies Giles[e]	–	–	7 (0.63)	0.7	59 (1.5)	2.0	2 (0.04)	0.04
An. peditaeniatus (Leicester)[f]	16 (1.93)	3.85	–	–	5 (0.13)	0.17	74 (1.66)	1.64
An. stephensi Liston[e]	–	–	2 (0.17)	0.2	1 (0.03)	0.03	2 (0.04)	0.04
An. subpictus Grassi[f]	–	–	119 (10.65)	11.9	346 (8.83)	11.53	25 (0.56)	0.55
Armigeres subalbatus (Coquillett)	8 (0.96)	1.92	2 (0.17)	0.2	3 (0.08)	0.1	31 (0.69)	0.68
Culex bitaeniorhynchus Giles[f]	5 (0.60)	1.20	2 (0.17)	0.2	2 (0.05)	0.07	27 (0.60)	0.6
Cx. fuscocephala Theobald[ef]	5 0.60	1.20	–	–	–	–	2 (0.04)	0.04
Cx. gelidus Theobald[ef]	30 (3.61)	7.21	–	–	–	–	37 (0.83)	0.82
Cx. perplexus Leicester	71 (8.54)	17.07	31 (2.77)	3.1	92 (2.35)	3.07	102 (2.28)	2.26
Cx. pseudovishnui Colless[ef]	3 (0.36)	0.72	4 (0.36)	0.4	8 (0.20)	0.27	146 (3.27)	3.24

(continued)

Table 7.4 (Continued)

| | Outdoor resting | | | | Indoor resting | | | |
| | BPD Hop Cage (9–13 h)[a] | | Hand catch (6–8 h)[b] | | Total catch (10-h)[c] | | Dusk collection (18–20 h)[d] | |
	Nos. (%)	No./h	Nos. (%)	No./h	Nos. (%)	No./h	Nos. (%)	No./h
Cx. quinquefasciatus Say[ef]	147 (17.69)	35.33	922 (82.47)	92.2	3281 (83.72)	109.37	2416 (54.10)	53.68
Cx. tritaeniorhynchus Giles[e]	512 (61.61)	123.07	8 (0.72)	0.8	46 (1.17)	1.53	778 (17.42)	17.29
Cx. vishnui Theobald[ef]	23 (2.77)	5.52	7 (0.63)	0.7	4 (0.10)	0.13	34 (0.76)	0.75
Mansonia annulifera (Theobald)[f]	–	–	–	–	2 (0.05)	0.07	3 (0.07)	0.06
Mn. uniformis (Theobald)[f]	–	–	–	–	–	–	7 (0.16)	0.15
Neomelaniconion lineatopennis (Ludlow)	7 (0.84)	1.68	5 (0.45)	0.5	70 (1.79)	2.33	675 (15.11)	15.0
Verralina indica (Edwards)	–	–	2 (0.17)	0.2	–	–	21 (0.47)	0.46
Grand total	831		1118		3919		4466	

[a]Mosquito collection in 4 h and 16 min using 1,280 hopping attempts at five hopping attempts/min in vegetation
[b]Mosquito collection in 10 h using aspirator tube and torch light for 10 min/village
[c]Mosquitoes collection in 30 h using Pyrethrum spray sheet method in one room/village at 30 min/room
[d]Mosquito collection in 45 h using aspirator tube and torch light during 6–8 PM for a period of 30–60 min/village [e]Medically important mosquito species
[f]Naturally infected with JE virus in India

Comparative efficacy of sampling tools used for collection of mosquitoes carried out at study sites of Saharanpur during 2005–2006 revealed that hop cage method was the most suitable surveillance tool for *Cx. tritaeniorhynchus* as it captured 7.1X, 80.4X and 153.8X more of female *Cx. tritaeniorhynchus* per hour than those by DC, TC and HC, respectively (Table 7.4).

7.3.5 Medical Importance of Mosquitoes of Saharanpur District

In India, JE virus has been isolated from 16 mosquito species viz., *Cx. bitaeniorhynchus, Cx. epidesmus, Cx. fuscocephala, Cx. gelidus, Cx. infula, Cx. psedovishnui, Cx. quinquefasciatus, Cx. tritaeniorhynchus, Cx. vishnui, Cx. whitmorei, An. barbirostris, An. peditaeniatus, An. subpictus, Ma. annulifera, Ma. indiana,* and *Ma. uniformis.* Among these species, maximum number of JE virus was isolated from *Cx. tritaeniorhynchus* (Philip et al. 2000; Dhanda and Kaul 1989). In the present study, 19 species were collected in dusk collection including 11 species which are reported to be naturally infected with JE virus in India (Fig. 7.4). *Culex* mosquitoes belonging to Cx. Vishnui subgroup, viz. *Cx. tritaeniorhynchus, Cx. pseudovishnui* and *Cx. vishnui* are important vectors of JE in India. In Saharanpur district, *Cx. tritaeniorhynchus* was the widely distributed JE vector species. It was the most abundant species in outdoors representing more than 61 % of total collection, while in DC the species accounted for 17.4 %, in TC for 1.2 % and in HC for 0.7 % of total female mosquito collection. In earlier studies also *Cx. tritaeniorhynchus* was the dominant species in outdoors contributing 95–100 % of the total mosquito collection in Karimnagar and Warrangal district, Andhra Pradesh (Das et al. 2004) and 77.4 % in Alleppey district, Kerala (Hiriyan et al. 2003). It is interesting to note that in areas where *Cx. tritaeniorhynchus* primarily rests outdoors, BPD hop cage method is the most suitable sampling tool for monitoring its density round the year. Earlier studies have shown that wild caught population of this species collected from different parts of the country were found repeatedly infected with JE virus. Of these, maximum number of virus positive mosquito pools of this species was from Tamil Nadu (67), followed by Kerala (7), Karnataka (3), Uttar Pradesh (1) and Haryana (1) (Carey et al. 1968; Philip Samuel et al. 2000; Pant et al. 1994; Das et al. 2005). In Saharanpur also, JE virus antigen was detected repeatedly in wild caught mosquitoes of *Cx. tritaeniorhynchus* and *Cx. vishnui* (included in the present study). *Chilodonella uncinata* (Ehrbg.), a ciliate, was found to cause chronic and fatal infection in mosquito larvae particularly *Cx. tritaeniorhynchus* and *Cx. pseudovishnui* in Sonipat district, Haryana state of India (Das 2003).

During July 2005, large numbers of *Cx. vishnui* males were collected from human dwellings close to a pond in Saharanpur district, while females were collected from vegetation. The species was not encountered after August in the area. In JE virus isolation study, it was second to *Cx. tritaeniorhynchus* and was found naturally infected with JE virus mainly from southern part of the country, viz. 23 JE virus isolations from Tamil Nadu, 4 from Karnataka and 1 from West Bengal (Philip Samuel et al. 2000). Larvae of *Cx. pseudovishnui* were frequently collected from burrow pits, pools and ponds in association with *Cx. tritaeniorhynchus* in Saharanpur district. However, the adult density of the species remained low in the area due to which it was not possible to screen the population for JE virus detection. So far, from India there were four JE virus isolations from this species, of which three were from Karnataka and one from Goa (Philip Samuel et al. 2000).

In Saharanpur district, during the month of September and October, immature stages of *Cx. fuscocephala* were abundant in large number of rain-fed ground pools in association with *Cx. tritaeniorhynchus*. JE virus antigen was detected from reared adults from these pools during the month of October (included later in this chapter). In India, JE virus (JEV) was isolated from wild caught specimens of *Cx. fuscocephala* and this species may have an important role in zoonotic cycle of transmission of JE (Philip Samuel et al. 2000). In Thailand, the species has been considered as an efficient vector of Japanese encephalitis (Muangman et al. 1972). Wild caught specimens of *Cx. fuscocephala* was reported to be infected with *Brugia malayi* in Sri Lanka (Carter 1948).

Among *Anopheles* species, *An. culicifacies* and *An. stephensi,* the well-known rural and urban malaria vector in India, were taken in dusk and indoor collections with negligible densities. These are zoophilic species and with such a low density ranging from 0.1 to 1.5 are insignificant as a vector of malaria in the area.

Cx. quinquefasciatus is the most common domestic species in India as it breeds in all types of water collections ranging from fresh to polluted like septic tanks. In Saharanpur district, it was found to be the dominant species in dusk and indoor collections and winter season in outdoor collections. *Cx. quinquefasciatus* has been incriminated as a principal vector of bancroftian filariasis in wide areas of the country. However, Saharanpur district is not an endemic area for filariasis. The females of this species are vicious biters of man from dusk to dawn while they also bite intensely on bovines during winter season in northern parts of the country. Although, it has been reported to be naturally infected with JE virus and was found capable of transmitting the virus experimentally (Philip Samuel et al. 2000), it is believed that *Cx. quinquefasciatus* is not an important vector of JE in India. The reason for its not being considered as a vector may be due to the fact that it rarely bites the amplifier host (pig) infected with JE virus (Sirivanakarn 1976).

All known vectors of JE encountered in the present study except *Cx. bitaeniorhynchus* are strongly zoophilic. Based on the abundance of vector mosquitoes encountered during transmission season of JE in Saharanpur and their virus antigen detection in culicine mosquitoes (described in more detail in Sect. 7.8), *Cx. tritaeniorhynchus* is considered as the primary vector of JE in Saharanpur district based on its high density and repeated detection of JE virus antigen. *Cx. fuscocephala* and *Cx. vishnui* may have secondary role in transmission of the disease in the area. Rest of the species mentioned in the systemic list of the district (Table 7.4) are of no medical importance in the area.

7.4   Comparative Seasonal Abundance of *Culex tritaeniorhynchus* from Affected and Unaffected Study Areas of Saharanpur District

7.4.1   *Proportion of Members of Culex vishnui Subgroup Taken in Affected and Unaffected Villages*

Table 7.5 shows percentage composition of members in *Culex vishnui* group collected in 2005–2006 in different types of collections, viz. resting collection: outdoor, indoor (total catch by pyrethrum space spray) and dusk collection undertaken in both JE affected and unaffected villages. Of the three members of *Culex vishnui* subgroup, *Cx. tritaeniorhynchus* was the most abundant species in all type of collections. However, proportion of this species was much higher in outdoor collection (96.4–98.8 %) and in dusk collection (90.2–97.6 %) than in indoor collection (71.4–83.7 %) in affected and unaffected villages, respectively. *Cx. vishnui* did not rest outdoors to any significant extent, while it did in indoor situation (16.3–28.6 %). *Cx. pseudovishnui* was the least abundant species in the study area.

Cx. tritaeniorhynchus was abundant in outdoors among secondary ground vegetation (in millets and mustard) and carpet type low level vegetation of fodder plant (berseem). They were also abundant in dusk collection, however, they were infrequent in indoor collection. Present observation differed from that of earlier findings from south India as well as from Singapore wherein it was stated that the species was very infrequent indoors and was relatively scare in outdoors also (Reuben 1971a, b, c; Colless 1959). In rural areas of Delhi Union Territory the species was found in very low densities only during monsoon and post-monsoon months (Menon and Rajagopalan 1976). Based on these observations it appears that the resting places of *Cx. tritaeniorhynchus* were not probably completely known in earlier studies. Millets, mustard and fodder plants were recognised as the ideal day time sources of this species in Saharanpur. *Cx. tritaeniorhynchus* was the most abundant species encountered 96–99 % of total mosquito collected from outdoor resting places in both JE affected and unaffected villages.

7.4.2   *Comparative Seasonal Abundance of Cx. tritaeniorhynchus, Primary JE Vector Species in Saharanpur District*

Analysis of data on adult mosquito collected from Saharanpur district during 2005–2006 indicate that among three species in *Culex vishnui* subgroup, *Cx. tritaeniorhynchus* mosquitoes were found throughout the year in outdoor situation; however, the adults of other two members, *Cx. vishnui* and *Cx. pseudovishnui* were found only in July and August and could not be collected from any

Table 7.5 Percentage of mosquitoes in *Culex vishnui* subgroup in different types collections in JE affected and unaffected study villages of Saharanpur district (2005–2006)

| Type of villages | | JE affected villages | | | | Unaffected Villages | | |
Species	Total no. of mosq.	Cx. tritaeniorhynchus	% Cx. vishnui	Cx. pseudovishnui	Total no. of mosq.	Cx. tritaeniorhynchus	% Cx. vishnui	Cx. pseudovishnui
Outdoor resting	329	96.35	3.34	0.31	243	98.78	1.23	–
Indoor resting	43	83.72	16.27	–	14	71.42	28.57	–
Dusk collection	589	90.15	4.75	5.09	808	97.64	0.62	1.73

Fig. 7.10 Seasonal
abundance of *Cx.*
tritaeniorhynchus in outdoor
and indoor collection in
affected study villages in
relation to transmission
season of JE in Saharanpur
district during 2005–2006

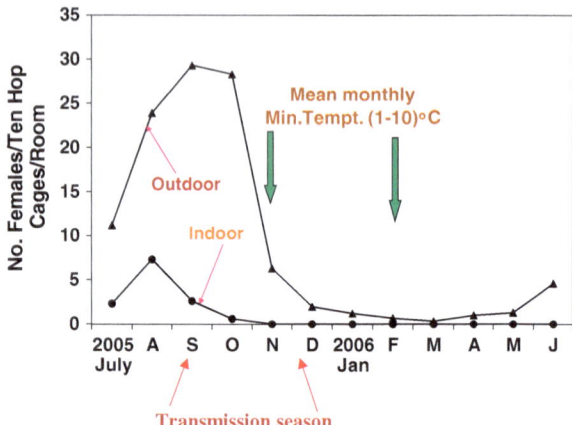

of the three resting situations during remaining period of the year. The density of
Cx. tritaeniorhynchus species in outdoor collection showed sudden increase from
July (11.2), coinciding with the beginning of the rains and remained elevated
till October (28.3) in affected villages during JE transmission season in the district (Figs. 7.5, 7.10). The population remained elevated during August–October
with peak 29.3 per ten hop cage density (PTHC) in September in affected villages. Suspected cases of JE started reporting from the district from 15 September
2005 after *Cx. tritaeniorhynchus* reached its population peak (PTHC 29.3)
in September in affected villages (Fig. 7.11). During the period, November–
December, the density of the species sharply dropped and remained at a very low
level (0.3–1.7) from December to May. In unaffected villages, the peak (33.7)
abundance of this species in outdoor collection was recorded earlier in August
(Fig. 7.11b). But thereafter, density of the species dropped rapidly and remained
fairly low ranging from 14.6 to 18.5 during major part of the transmission season
(September–November).

Under indoor situation, the per room density (PRD) of *Cx. tritaeniorhynchus*
was poor, ranged from 0.6 to 7.3 and 0.5 to 3.0 in affected and unaffected villages,
respectively. From July to October 2005, indoor resting density by total catch
method of *Cx. tritaeniorhynchus* in human dwellings in affected and unaffected
villages were poor. In contrast, during the same period, 5–47 times and 9–40
times more mosquitoes of this species were found to rest in outdoor vegetation in
affected and unaffected villages, respectively (Fig. 7.11).

Cx. tritaeniorhynchus abundance in dusk collections was nil for 3 months
(November to January), remained low for next 6 months (February–July) ranging
from 0 to 9.5 per man hour (PMH) and increased rapidly during August–October
in the study villages. The peak PMH density of the species in dusk collection
in September was 222 and 321 in affected and unaffected villages, respectively
(Fig. 7.12a). During peak transmission period (mid-November) dusk collection
resulted nil for *Cx. tritaeniorhynchus* mosquitoes as winter season already started
in the district and females of vector species had stopped taking blood meal.

Fig. 7.11 Comparative monthly abundance of *Cx. tritaeniorhynchus* in outdoor and indoor collection in Saharanpur district during (2005–2006). **a** JE affected study villages. **b** Unaffected study villages

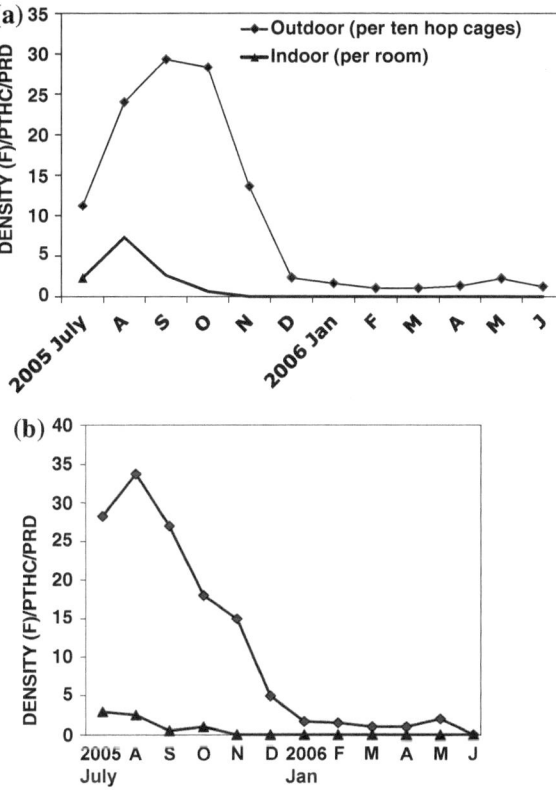

BPD hop cage method was the most suitable tool for monitoring *Cx. tritaenio-rhynchus* abundance round the year under outdoor situation in Saharanpur district. During outbreak/epidemic situation also, hop cage method was used effectively to estimate JE vector abundance in Northern and Southern India (Gupta et al. 2005; Das et al. 2004). In the present study, the species was found with 5–47 times and 9–40 times higher densities in outdoors than that in indoors in affected and unaffected villages, respectively. Indoor collection following total catch method, therefore, is not a reliable tool for JE vector surveillance. In southern India, JE vectors remained physiologically active round the year due to the absence of extreme winter season (Kanojia 2007). Hence, dusk index, the average number of parous females collected per man hour in dusk collection was used to monitor JE vector abundance in southern India (Mani et al. 1991; Reuben et al. 1992; Gajanana et al. 1997; Gajanana and Arunachalam 1998; Philip Samuel et al. 1998; Thenmozhi et al. 2001). But in Northern India, dusk index can be used to monitor JE vector abundance only for 6 months from May to October (Fig. 7.12) and not during November when maximum number of suspected cases of JE were reported in the district, in 2005.

Fig. 7.12 **a** Seasonal
abundance of *Cx.*
tritaeniorhynchus in dusk
collection in JE affected and
unaffected study villages.
b Transmission season of JE
in Saharanpur district

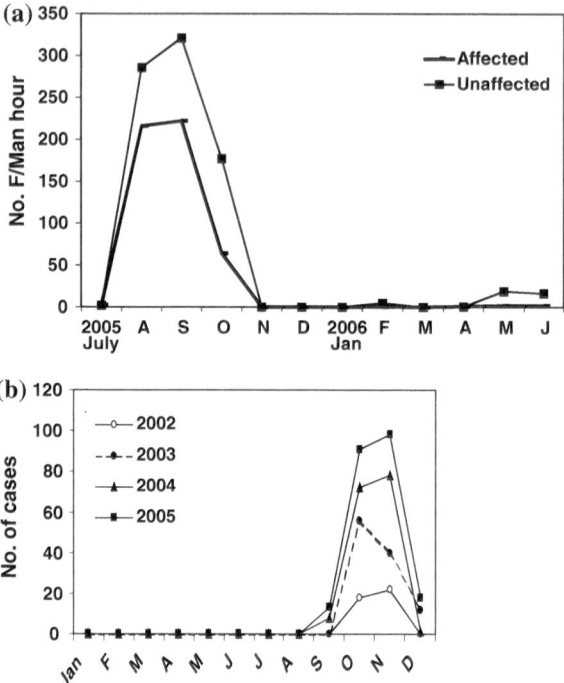

Although there was no direct correlation between rain fall pattern and progres-
sion of JE cases in Saharanpur district, a significant positive correlation was found
between rain fall pattern and abundance of *Cx. tritaeniorhynchus* mosquitoes in
the district. In affected villages, *Cx. tritaeniorhynchus* showed a slow but gradual
increase in density from March till June with a very small peak in May followed
by a sharp increase in vector density during the next 5 months (July–November)
with a tall peak in September (Fig 7.11a). However, in unaffected study villages
the tall peak reached earlier (August) followed by substantial decrease in vector
density during the transmission season. Bimodal pattern of population density was
reported in *Cx. vishnui* group in Gorakhpur district (UP state of India) from where
large number of suspected cases of JE is being reported regularly since 1978
(Kanojia et al. 2003).

The period of 2005 outbreak in Saharanpur district was from September to
December with peak in November (Fig. 7.5). Indoor densities of *Cx. tritaeniorhynchus* were poor and restricted only for 4 months of the year (July–October).
A team visiting the district in November–may report nil density of *Cx. tritaeniorhynchus* in indoor collection as well as dusk collection. With the result, lot of
confusion will be created and the disease cannot be established based on entomological evidence. Under the situation, adult mosquito survey in outdoor vegetation
using BPD hop cage method will be useful. However, after outbreak of human
cases in an area, visiting team would always be at disadvantage, due to usual time

Fig. 7.13 Resting habitats of *Cx. tritaeniorhynchus* in Saharanpur district (2005)

lag, of losing the vital information on the population peak of vector species preceding occurrence of human case. Information so far available on JE vector abundance from Northern India is all based on outbreak investigations involving indoor resting collection, hence is bound to be erroneous. Thus, monitoring of vector (mosquito) density is a prerequisite for developing an effective vector control measures to prevent transmission of the disease in JE endemic districts/states of Northern India.

7.5 Shifting Pattern in the Resting Habit of *Cx. tritaeniorhynchus* in Saharanpur District

Culex tritaeniorhynchus rested predominantly in millets, mustard and berseem but rarely in hyacinth pond in Saharanpur. Millets supported resting population of JE vectors from June to October in both affected and unaffected villages of Saharanpur district. Paddy was never found to support day resting habitat for JE vectors. By the end of October, both millet and paddy were harvested and mustard was available in the area for 2 months (November–December). *Cx. tritaeniorhynchus* mosquitoes shifted to the grassy secondary vegetation under the mustard crop (Fig. 7.13). In affected villages, the population abundance of the species rapidly declined from 28.3 PTHC in millet in October (Fig. 7.14a) to 13.6 PTHC (November) and 1.9 (December) in mustard crop (Fig. 7.14b). The density of *Cx. tritaeniorhynchus* in unaffected villages was 15 PTHC in millet (October) and 8.0 PTHC in mustard (December). This decrease in density in November–December was due to sudden drop in mean monthly minimum temperature that ranged from

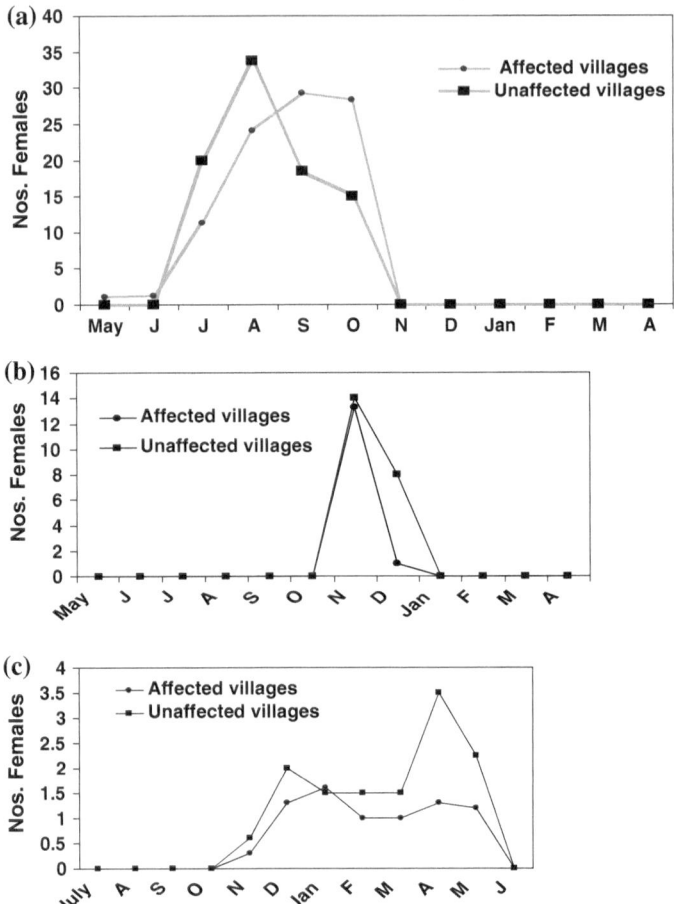

Fig. 7.14 Shifting behaviour of outdoor resting habit of *Cx. tritaeniorhynchus* mosquitoes in affected and unaffected villages of Saharanpur district. **a** Millet, **b** Mustard, **c** Berseem

14.4 in October to 8.4, 4.5 and 5.4 in November, December and January, respectively. Another crop, berseem was available in the district from December till May. Mustard was harvested in December and *Cx. tritaeniorhynchus* mosquitoes shifted to berseem in December. Wild caught *Cx. tritaeniorhynchus* mosquitoes during November were mainly unfed. As the night temperature dropped in the district, females stopped taking blood meal and passed the cold winter months resting first in mustard then in berseem.

Male population totally disappeared for next 3 months (December–February) as the females remained unfed for 3 months, viz. November to January (overwintering in *Cx. tritaeniorhynchus*). From mid-February certain percentage of outdoor resting as well as dusk collections yielded few *Cx. tritaeniorhynchus* fed females (Fig. 7.15). Due to gradual rise in night temperature, overwintering females started

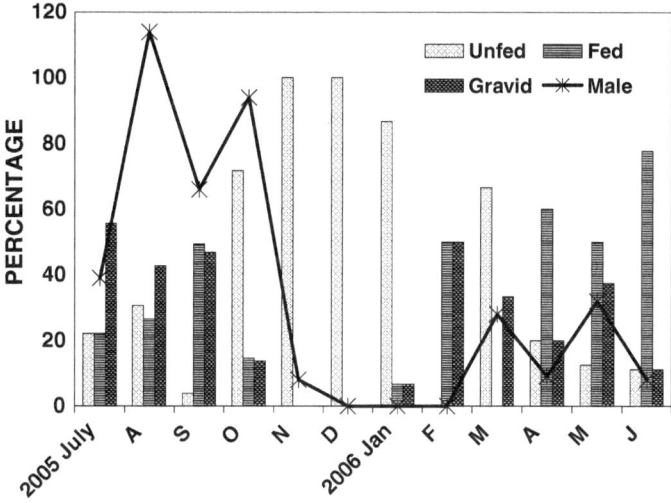

Fig. 7.15 Males and percentage of females (Unfed, Fed and Gravid) *Cx. tritaeniorhynchus* mosquitoes collected from outdoor vegetation of Saharanpur district

taking blood feed and the population gradually increased from March onwards. However, so far *Cx. quinquefasciatus* never found to overwinter in UP, Haryana and Delhi. *Cx. tritaeniorhynchus* continued to rest in berseem till May and PTHC density of the species in the study villages ranged from 1.5 to 2.2 (Fig. 7.14C). The prolonged low density of the species was due to limited breeding sites like temporary grassy pools around hand pump and ponds. The other two vectors species, *Cx. vishnui* and *Cx. pseudovishnui* were also found to rest in millet plants only during July and August and were not detected in any other vegetation. Paddy field never supported mosquito resting due to the presence of water and absence of secondary grassy vegetation. Wild vegetation did not supported mosquito resting in the study area, exact reason was not fully understood. Sugar cane did support grassy vegetation underneath in some months and may support a fraction of resting population of mosquitoes in those months. However, hop cage could not be used for collecting mosquitoes from sugar cane as cotton hop cage got damaged due to sharp leaves of sugar cane.

A shift in the outdoor resting behaviour of *Cx. tritaeniorhynchus* was noticed in accordance with the seasonal cropping pattern in study areas. As the members of *Cx. vishnui* group (*Cx. tritaeniorhynchus*, *Cx. vishnui* and *Cx. pseudovishnui*) show a succession in breeding behaviour in rice fields (Reuben 1971a, b, c), the adults of *Cx. tritaeniorhynchus* also found to change its resting habitat in accordance with the local agriculture practice. Millet supported very high population of *Cx. tritaeniorhynchus* (July–October), mustard supported mainly the overwintering population and berseem the lean period of population abundance of this species. During September 2005, one suspected case of JE was reported from one of the study village, Manak Mau and in the present study JE virus antigen was detected in *Cx. tritaeniorhynchus* (July and August) collected from millet crop of the village. Another study village, Chilkana reported two JE cases during October 2005 and JE virus activity was detected in *Cx. tritaeniorhynchus* (August and October) collected from millet. Therefore, millet was found to be the ecologically dangerous niche that required immediate attention to interrupt disease transmission. Thus, best way to get large number of adult females of *Cx. tritaeniorhynchus* in Northern India for estimation of vector density as well as detection of JE virus infection is to collect mosquitoes using BPD hop cage method from millets during July–October and mustard in November.

7.6 Feeding Behaviour of *Cx. tritaeniorhynchus* in Saharanpur District

Female specimens of *Cx. tritaeniorhynchus* collected from outdoor situations during April–October were in all stages of gonotrophic cycle, viz. unfed, fed and gravid. From mid-November onwards, the night temperature in the area started falling and remained low (below 10 °C) till mid-February. *Cx. tritaeniorhynchus* females suspended all its physiological activities during later part of November and remained quiescent in their outdoor resting locations in mustard and fodder plants till around mid-February. The male population of *Cx. tritaeniorhynchus* disappeared during December to February from the area (Fig. 7.15), this indicates cessation of breeding of the species during this period. Similar trend in population behaviour of the species was noted in other JE endemic areas of North India like in Delhi and Sonipat and Karnal district of Haryana (Chap. 5, Chap. 6).

A total of 207 mosquitoes, mainly of *Cx. tritaeniorhynchus* (87 %) were tested for their blood meal source, which revealed that *Cx. tritaeniorhynchus* predominantly (87.6 %) fed on cattle, while only 0.6 % fed on humans, and 11.8 % did not show either of the two. Among other species of mosquitoes the feeding was mainly on cattle: *Cx. gelidus* (100 %), *Cx. perplexus* (94 %), *Cx. bitaeniorhynchus* (50 %). However, *An. culicifacies* showed partly zoophilic behaviour as only 22 % were found to feed on bovine blood. In the unaffected study villages, population peak of *Cx. tritaeniorhynchus* was higher and occurred earlier (August) and JE virus infection was detected in a pool made out of outdoor collection (millet). Thereafter, vector abundance declined and no human case was reported from these villages. In presence of abundant cattle population (cattle: human = 2.3–2.5:1.0) with economically poor local residents in unaffected village (Pilakhni and Halalpur), the predominantly zoophilic vector might have failed to transmit the virus to human. This indicates zooprophylaxis is likely to have beneficial effect on disease transmission. However, the percentage of *Cx. tritaeniorhynchus* (0.6 %) fed on human population in Saharanpur district was higher than those reported earlier from North Arcot district (0.2 %) and lower than those from Madurai and South Arcot district (2.1–4.1 %) in southern India (Christopher and Reuben 1971;

Reuben et al. 1992). This feeding behaviour during active transmission period has epidemiological implications and may be exploited to minimise man–vector contact by appropriate vector management practices. *Cx. tritaeniorhynchus* was found to be dusk/early hours of night feeder in these villages and rarely fed on human beings under high-density level. Therefore, use of mosquito net is unlikely to have any beneficial effect as far as control/management of Japanese encephalitis is concerned.

7.7 Brick Kiln Industry and Its Role in JE Vector Breeding Potential in Saharanpur District

Paddy fields, permanent ponds, hyacinth marshes and extensive burrow pits near brick kiln located in affected and unaffected villages of Saharanpur were sampled at monthly interval from July 2005 to June 2006 and larval density of culicines were assessed. As shown in (Fig. 7.16) paddy fields were the least preferred breeding habitat of *Cx. tritaeniorhynchus,* per dip larval density ranging from 1–8, followed by pond (7–25). Rain water pools around brick kiln supported maximum (0.3–80 per dip) JE vector breeding (Fig. 7.16c). *Cx. fuscocephala* was found breeding profusely in rain water pools near brick kiln along with *Cx. tritaeniorhynchus* during the month of October (Fig. 7.17). During the study period, paddy fields were not found to be water logged for a longer period and larval sample collected usually comprised younger stages only. Large number of temporary pools (excavated land around each brick kiln) were created after monsoon rain in July and these with luxuriant aquatic vegetation provided extensive breeding of JE vectors till end of October. These and hyacinth ponds contributed nearly 95 % of *Cx. tritaeniorhynchus* breeding mainly in monsoon months (July–October) and remaining 5 % was contributed by paddy fields from mid-July to early October (Fig. 7.18). During winter months (December–February), no larvae of this species was detected in the study area. Scanty breeding of *Cx. vishnui* and *Cx. pseudovishnui* was detected in hoof prints and margin of permanent ponds used for rearing edible fish during July and November respectively.

Of the five study sites, per capita cattle population was lower in Chilkana and Manak Mau as evidenced by the human: cattle: pig ratio: Chilkana (20,000: 2,500: 40), Manak Mau (15,000: 3,000: 50), Bhudhakhera Ahir (4,000: 1,000: 0), Halalpur (7,000: 3,000: 30) and Pilakhni (5,000: 2,000: 100). *Cx. tritaeniorhynchus* mosquitoes were found to be strongly zoophilic, abundant and widely distributed in all the study sites. Chilkana and Manak Mau were bigger villages with economically better off population remained at the risk of disease incidence as far as JE is concerned due to lower per capita cattle population and presence of several brick kiln industries at the peripheral areas in these villages. In contrast, with mainly economically poor population, Halalpur and Pilakhni were at the lower risk of disease incidence due to higher per capita cattle population and absence of brick kiln industries in these villages. Mud houses were predominant at Halalpur

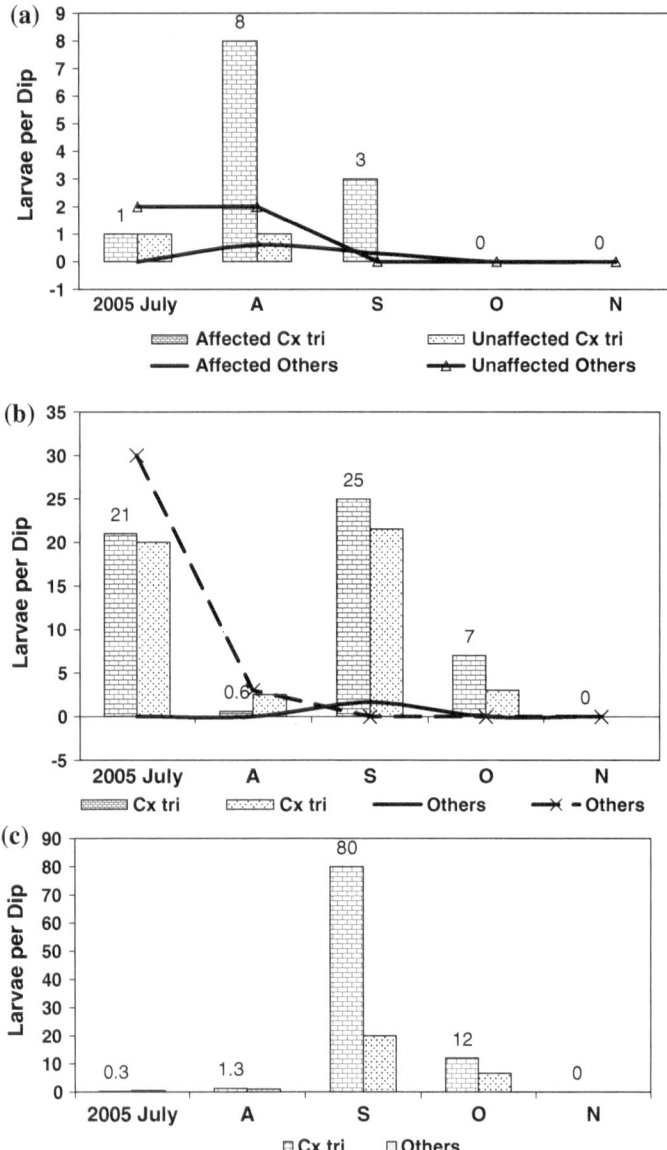

Fig. 7.16 Comparative breeding preference of *Cx. tritaeniorhynchus* and other mosquito species in Affected and Unaffected study villages of Saharanpur district. **a** Paddy fields. **b** Ponds. **c** Extensive rain-fed marshy area around brick factory

village and burrow pits appears to contribute immensely to highest prevalence of *Cx. tritaeniorhynchus*. Although economically better off, Budhakhera Ahir also remained at the lower risk due to absence of brick kiln industry and relatively low abundance of vector species (*Cx. tritaeniorhynchus*) which may be due to the

Fig. 7.17 Role of rain water pools around brick factory of Saharanpur district in breeding JE vector (2005). **a** Manak Mau, June (Dry). **b** Chilkana (August), *Cx. tritaeniorhynchus* (20 per dip). **c** Chilkana (Oct.), *Cx. tritaeniorhynchus* + *Cx. fuscocephala* (80 per dip). **d** Chilkana (Nov.) *Cx. tritaeniorhynchus* (2 per dip)

Fig. 7.18 Breeding potential of *Cx. tritaeniorhynchus* in Saharanpur district. **a** Paddy fields. **b** Pond and Marshes. **c** Rain water pools near Brick kiln factory

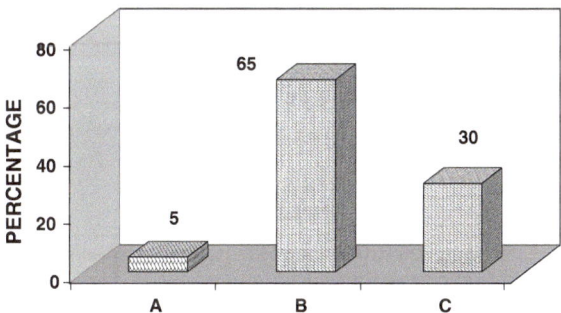

presence of ciliate parasite (*Chilodonella uncinata*) acting as natural biolarvicide as detected elsewhere (Das 2003).

In South India, paddy fields were the major breeding source of *Cx. tritaenio-rhynchus* mosquitoes (Reuben 1971a, b, c), similar pattern was noted in Karnal District of Haryana and Gorakhpur district of Terai region of UP. Paddy fields in these areas remain water logged during monsoon months (July–September) and widespread breeding of several mosquito species, including *Cx. tritaeniorhynchus* was found. In contrast, in Saharanpur district paddy fields never found to hold water for more than a week, hence were not the ideal breeding source for this species. Similar observation was reported by (La Casse and Yamaguti 1955) in related studies from Japan.

Table 7.6 Wild caught adult mosquitoes (outdoor/indoor/dusk) and those reared from larvae [adult (L)] of four JE vector species collected in Saharanpur district (2005–2006) tested for JE virus infection by ELISA[a]

Species/Type of Collection	Affected villages			Unaffected villages		
	Males[b]	Females[b]	Total[b]	Males[b]	Females[b]	Total[b]
July–December 2005 (Group I)						
Cx. tritaeniorhynchus						
Outdoor	4/8 (375)	1/5 (69)	**5/13** (444)	1/4 (237)	0/5 (77)	**1/9** (314)
Indoor	–	–	–	–	0/1 (2)	0/1 (2)
Dusk	0/1 (33)	0/4 (148)	0/5 (181)	–	0/2 (57)	0/2 (57)
Adults (L)	0/2 (14)	–	0/2 (14)	–	0/2 (21)	0/2 (21)
Cx. vishnui						
Outdoor	–	–	-	0/2 (98)	0/1 (5)	0/3 (103)
Indoor	1/1 (100)	–	**1/1** (100)	–	–	–
Dusk	–	0/1 (5)	0/1 (5)	–	0/1 (27)	0/1 (27)
Cx. fuscocephala						
Adults (L)	–	1/1 (16)	**1/1** (16)	–	–	–
Cx. gelidus						
Outdoor	–	0/1 (12)	0/1 (12)	–	–	–
January–June 2006 (Group II)						
Cx. tritaeniorhynchus						
Outdoor	0/1 (7)	0/1 (4)	0/2 (11)	–	0/1 (2)	0/1 (2)

[a]ELISA using detector antibody MAB 6B6C-1, cut-off OD 0.060
[b]Number of pools positive/number of pools (number of mosquitoes) tested

7.8 Detection of Natural Vertical Transmission of JE Virus in Vector Mosquitoes of Saharanpur District

Altogether, a total of 1,309 mosquitoes belonging to two groups in 45 pools: *Cx. tritaeniorhynchus*—1,046 mosquitoes (37 pools), *Cx. vishnui*—235 (6 pools), *Cx. fuscocephala*—16 (1 pool) and *Cx. gelidus*—12 (1 pool) were screened for JE virus infection by ELISA (Gajanana et al. 1995). 8 out of 42 (19 %) pools in (Group I) had OD values above 0.060 (mean + 4 SD). 3 pools in Group II were found to be negative by ELISA. JE virus antigen was detected in 3 out of 4 culicine species, viz. *Cx. tritaeniorhynchus, Cx. vishnui* and *Cx. fuscocephala*. The number of vector mosquitoes collected in affected and unaffected villages during the study period from Saharanpur district tested for JE virus infection by ELISA is shown in Table 7.6. Of the 8 positive pools, 6 (75 %) were from *Cx. tritaeniorhynchus* pools and mosquitoes

Table 7.7 JE virus antigen detection from mosquitoes in affected and unaffected villages in Saharanpur district (2005–2006)

Species	JE affected study villages			Unaffected study villages		
	No. pools tested[a]	No. positive	MIR[b]	No. pools tested[a]	No. positive	MIR[b]
Cx. tritaeniorhynchus	22 (650)	5	7.69	15 (396)	1	2.52
Cx. vishnui	2 (105)	1	9.52	4 (130)	0	0
Cx. fuscocephala	1 (16)	1	62.5	–	–	–
Cx. gelidus	1 (12)	0	0	–	–	–
Total	26 (783)	7	8.94	19 (526)	1	1.90

[a]Number of mosquitoes in parenthesis
[b]MIR, Minimum infection rate/1000 specimen tested

of 5 positive pools (4 males and 1 female) of this species were collected from outdoors (millets) in affected villages during the month of July (1 male pool), August (2 male pools and 1 female pool) and October (1 male pool). Mosquitoes of the sixth positive pool of Cx. tritaeniorhynchus was also collected from millet crop in unaffected villages in August. Out of six pools of Cx. vishnui tested, mosquitoes of the positive pool were males collected from human dwelling in affected villages in July. Similarly, mosquitoes in the single pool of Cx. fuscocephala tested positive were females reared from larvae collected from extensive rain water burrow pits in affected villages in October. Of the nine pools (seven Cx. tritaeniorhynchus and two Cx. vishnui) made out of dusk collections, none was found to be positive for JE virus. While out of 28 mosquito pools (25 Cx. tritaeniorhynchus and three Cx. vishnui) made from outdoor collection, 6 (21.4 %) pools were found to be positive for JE virus infection.

JE virus antigen detected from mosquitoes from affected and unaffected villages are presented in Table 7.7. The overall minimum infection rate (MIR) per thousand mosquitoes tested for Cx. tritaeniorhynchus were 7.69 and 2.52 in affected and unaffected, respectively. In affected villages, MIR per thousand mosquitoes in Cx. vishnui was 9.52 and that of Cx. fuscocephala was 62.5. During September 2005, one suspected case of JE was reported from one of the affected village Manak Mau and in the present study JE virus activity was detected in two vector mosquitoes: Cx. tritaeniorhynchus (July and August 2005) and Cx. vishnui (July 2005) from the same village. Another affected village, Chilkana reported two JE cases during October 2005. In the present study JE virus activity was detected in two vectors: Cx. tritaeniorhynchus (August and October 2005) and Cx. fuscocephala (October 2005). In view of substantially high population density of Cx. tritaeniorhynchus during transmission season and maximum number of JE virus antigen detected from affected villages, this species is likely to be the primary vector of JE virus in Saharanpur district.

Pant et al. (1994) first time isolated JE virus from wild caught Cx tritaeniorhynchus mosquitoes in Gorakhpur, UP, North India. Vertical transmission

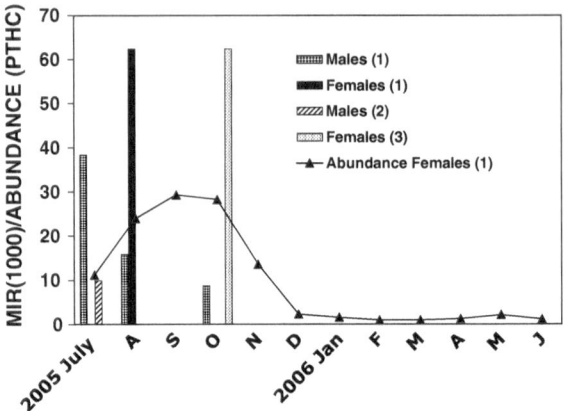

Fig. 7.19 Seasonal variations in vertical transmission of JE virus (*bars*) in males and females of JE vectors and in population abundance of *Cx. tritaeniorhynchus* under outdoor situation (*line*) in affected villages of Saharanpur district. *1 Cx. tritaeniorhynchus, 2 Cx. vishnui* and *3 Cx. fuscocephala*

of the JE virus was established in Saharanpur in three mosquito species (*Cx. tritaeniorhynchus*, *Cx. vishnui* and *Cx. fuscocephala*) for the first time in northern India. Earlier studies elsewhere have also demonstrated vertical transmission in these mosquitoes in Southern India (Rosen et al. 1978; Dhanda et al. 1989; Thenmozhi et al. 2001; Arunachalam et al. 2002). In Saharanpur district, the local residents are mostly Muslims and pig population is very poor in the area. Vertical transmission of JE virus in mosquitoes might be the main mechanism for the maintenance of JE virus in nature in area.

7.9 Seasonal Variations in Minimum Infection Rates of JE Virus in Vector Mosquitoes in Relation to Seasonal Abundance of *Cx. tritaeniorhynchus* in Saharanpur District

Seasonal variations in natural vertical transmission of JE virus in vector mosquitoes in relation to vector (*Cx. tritaeniorhynchus*) abundance (female/Ten hop cages) in affected and unaffected villages are shown in Figs. 7.19 and 7.20 respectively. In July 2005, there were two positive pools (one each of *Cx. tritaeniorhynchus* and *Cx. vishnui*). MIR per thousand male mosquitoes of the former species was found to be very high (38.46) and that of *Cx. vishnui* was 10.0. PTHC density of *Cx. tritaeniorhynchus* for the corresponding month was 11.2 in the affected villages. In August, MIR of *Cx. tritaeniorhynchus* females was very high (62.5) and that of male was 15.87. Similarly, PTHC density of *Cx. tritaeniorhynchus* was 24 in affected villages in August. In October, MIR of *Cx. fuscocephala* females (reared adults) was very high (62.5) and that of *Cx. tritaeniorhynchus* males was

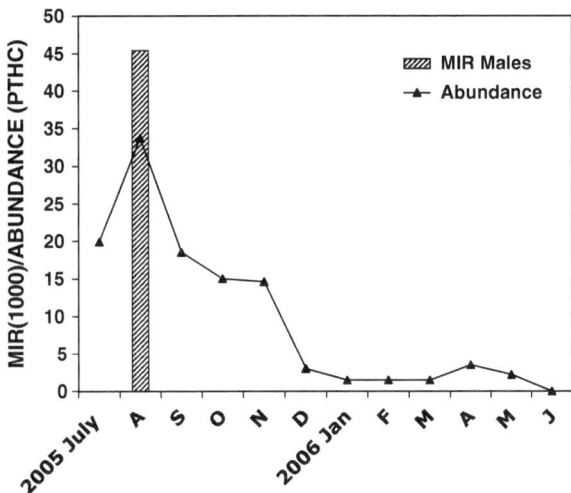

Fig. 7.20 Seasonal variation in vertical transmission of JE virus in males (*bar*) and in population abundance of females (*line*) of *Cx. tritaeniorhynchus* under outdoor situation in unaffected villages of Saharanpur district

low (8.77) and PTHC density of *Cx. tritaeniorhynchus* was 28.3 in the affected villages (Fig. 7.19). In unaffected villages, JE virus infection was detected only in the month of August in *Cx. tritaeniorhynchus*. MIR of *Cx. tritaeniorhynchus* males was 45.45 and its abundance in females under outdoor situation in these villages was 33.75 PTHC (Fig. 7.20). Based on the elevated vector density which was maintained during the monsoon months and with repeated detection of JE virus infection, *Cx. tritaeniorhynchus* was considered responsible for causing seasonal viral encephalitis in affected villages of Saharanpur district.

MIR is an important indicator of risk of JE transmission in an area. For *Cx. tritaeniorhynchus* the MIR was found to be much higher (7.69) in JE affected villages of Saharanpur district (Table 7.7) than in Anuradhapura district of Sri Lanka where the MIR was 0.12 (Peiris et al. 1992) and 0.28 from South Arcot district, South India (Gajanana et al. 1997). In the neighbouring JE endemic Karnal district of Haryana, MIR of *Cx. tritaeniorhynchus* was 6.06 in October and a total of 86 suspected cases of JE were reported from the district (Das et al. 2005). When JE occurrence was analysed together with detection of JE virus infection in JE vectors, it was found that vertical transmission of JE virus occurred in two species (*Cx. tritaeniorhynchus* and *Cx. vishnui*) 2 months prior to reporting of human cases in Saharanpur district. This is an early warning signal for initiating integrated vector control measures to prevent JE outbreak.

7.10 JE Vector/Virus Surveillance

Keeping in view the above findings, following vector/virus surveillance methods are suggested for Saharanpur district:

7.10.1 JE Vector Surveillance

7.10.1.1 Adult Mosquito Survey in Outdoors

Mosquito survey is to be carried out using BPD hop cage method in following sequence: Millet (June–October), Mustard (November) and Berseem (December–May).

7.10.1.2 Dusk Time Resting Mosquito Collection

Mosquito survey is to be carried out in and around cattle sheds/human dwellings using aspirator tube and torch light from May to October.

7.10.2 JE Virus Surveillance in Mosquitoes

Wild caught unfed culicine mosquitoes collected from millets (June–October) and mustard (November) need to be sorted into pools species wise and sex wise, each containing about 50 specimens according to species, sex, place and date of collection. These pools are to be kept as dry specimens at room temperature (25 ± 2 °C). These dry mosquito pools may be transported without cold chain facility to National Centre for Disease Control (erstwhile NICD), Delhi or Centre for Research in Medical Entomology (CRME), Madurai, Tamil Nadu for detection of JE virus infection in vector mosquito species.

7.11 JE Vector Control/Management Strategies for Saharanpur District

Japanese encephalitis continues to be a National problem, not because of poverty but because of lack of awareness. JE control is possible through protection of human and pig, and reduction of adult vector (mosquito) population. Control strategies can be drawn and implement to protect human and pig, and to reduce adult mosquito population below a thresh hold level. However, each method has its own limitation and it is not practical to depend on any single method. Therefore, an integrated approach is suggested for prevention/control of Japanese encephalitis.

7.11.1 Protection of Human

Health education to the general public regarding the method of transmission of disease and its prevention needs to be imparted. Using mass media, public should be advised to use of bed nets, door screen, impregnated curtains: (i) Use of insecticide impregnated curtains in the district from June to November to repel vector mosquitoes for reduction of man–mosquito contact which is essential to interrupt JE virus transmission, (ii) Burning of herbal materials like dry leaves (neem, etc.) during evening hours (1 h before and after dusk) for 4 months (July–October) to minimise man–vector contact.

Long-term vector control measure requires coordinated efforts by Panchayati Raj, Municipal/Urban Development Administration and Agricultural Department for environmental sanitation: (i) Land filling of low lying areas near each and every brick kiln during December–May to prevent rain water stagnation in monsoon months, (ii) Deweeding of ponds and marshes during December–May to prevent JE vector breeding.

7.11.2 *Reduction of Adult Vector Population*

In order to target exophilic vector species (*Cx. tritaeniorhynchus* mosquitoes) following intervention measures are recommended: (i) Malathion fogging operation may be undertaken in millet crops during July–October and mustard during November, (ii) Simultaneously, indoor resting population of *Cx. tritaeniorhynchus* in human dwelling may be controlled by malathion fogging operation during August–October. Vector species need to be targeted in both indoor and outdoor habitats as possibility of vector mosquito being infected is more based on the finding that vertical (transovarian) transmission has been found to be the main mechanism of JE virus maintenance in nature in Saharanpur district, (ii) Application of a suitable biolarvicide having good biological properties (capable to recycle in aquatic habitat, disperse in the environment via transovarian transmission, tolerant to desiccation and ultraviolet light) in stagnated rain water pools around brick kiln and ponds, (iii) Stocking and maintenance of larvivorous fishes in permanent ponds with water hyacinth problem to eliminate/reduce mosquito breeding, and (iv) Physical control of weed: water hyacinth problem is comparatively less in Saharanpur district and these can easily be physically removed to eliminate JE vector breeding from weed-infested area.

7.11.3 *Measures to Protect Pigs*

In order to reduce vector–amplifier host (pig) contact, following measures are suggested: (i) Piggeries should be constructed in a proper way or kept well protected from the mosquitoes, (ii) Appropriate insecticidal spray in the pig sheds to reduce the JE vector density and to break the transmission cycle.

7.12 JE Vaccination Programme in Saharanpur District

The community of Saharanpur cooperated well with the mass JE vaccination campaign and a total of 9.28 lakhs children out of the target of 10.6 lakhs were immunised during May and June 2007. Many children with acute viral encephalitis symptoms were admitted in Saharanpur District Hospital in 2007 who gave a history of receiving JE were vaccinated with a single dose of Chinese vaccine SA14-14-2 strain vaccination (Panwar and Kumar 2008).

Saharanpur district was considered as an 'endemic' area where children (<15 years) alone are vulnerable to JEV infection (Sabesan et al. 2008), therefore the district need to be included in future mass vaccination programme of Japanese encephalitis to be carried out in the country.

References

Arunachalam N, Philip Samuel P, Hiriyan J, Thenmozhi V et al (2002) Vertical transmission of Japanese virus in *Mansonia* species, in an epidemic-prone area of southern India. Ann Trop Med Parasitol 96:419–420

Carey DE, Reuben R, Myres RM, Pavri KM et al (1968) Japanese encephalitis studies in Vellore, South India. I: virus isolation from mosquitoes. Indian J Med Res 56:1309–1318

Carter HF (1948) Records of filarial infections in mosquitoes in Ceylon. Ann Trop Med Parasitol 42:312–321

Christophers S, Reuben R (1971) Studies of the mosquitoes of North Arcot District, Madras, South India. Part 4: host preference as shown by precipitin tests. J Med Entomol 8:314–318

Colless DH (1959) Notes on the culicine mosquitoes of Singapore. VII.—host preferences in relation to the transmission of disease. Ann Trop Med Parasitol 53:259–267

Das BP (2000) A new technique for sampling outdoor resting population of *Culex tritaeniorhynchus*, vector of Japanese encephalitis. In: Fourteenth national congress of parasitology, New Delhi, Abstr. No. PS-15, pp 133–134, 23–26 April 2000

Das BP (2003) *Chilodonella uncinata*—a protozoa pathogenic to mosquito larvae. Curr Sci 85:483–489

Das BP (2009) BPD hop cage method—a new device of collecting mosquitoes for effective JE vector surveillance. Invent Intell 44:24–25

Das BP, Lal S, Saxena VK (2004) Outdoor resting preference of *Culex tritaeniorhynchus*, vector of Japanese encephalitis in Warangal and Karim Nagar district, Andhra Padesh. J Vector Borne Dis 41:32–36

Das BP, Rajagopal R, Akiyama J (1990) Pictorial key to the species of Indian Anopheline mosquitoes. J Pure Appl Zool 2:131–162

Das BP, Sharma SN, Kabilan L, Lal S et al (2005) First time detection of Japanese encephalitis virus antigen in dry and unpreserved *Culex tritaeniorhynchus* mosquitoes Giles, 1901, from Karnal district of Haryana state of India. J Commun Dis 37:131–133

Dhanda V, Mouyra DT, Mishra AC, Ilkal MA et al (1989) Japanese encephalitis virus infection in mosquitoes reared from field collected immatures and in wild caught males. Am J Trop Med Hyg 41:732–736

Gajanana A, Arunachalam N (1998) Mosquito transmitted flavivirus infections in India. In: Goel SC (ed)Advances in Medical Entomology & Human Welfare (Supplement I.). Uttar Pradesh Zoological Society, Muzaffarnagar, pp 89–100

Gajanana A, Rajendran R, Thenmozhi V, Philip Samuel P et al (1995) Comparative evaluation of bioassay and ELISA for detection of Japanese encephalitis virus in field collected mosquitoes. Southeast Asian J Trop Med Public Health 26:91–97

Gajanana A, Rajendran R, Philip Samuel P, Thenmozhi V et al (1997) Japanese encephalitis in south Arcot district, Tamil Nadu, India: A three year longitudinal study of vector abundance and vector infection frequency. J Med Entomol 34:651–659

Gupta N, Hossain S, Lal R, Das BP et al (2005) Epidemiological profile of Japanese encephalitis outbreak in Gorakhpur, U.P. in 2004. J Commun Dis 37:145–149

Hiriyan J, Arunachalam N, Philip Samuel P et al (2003) Studies on a mosquito fauna in a Japanese encephalitis prone area in Kerala, India. Entomon 28:139–146

Kanojia PC (2007) Ecological study on mosquito vectors of Japanese encephalitis virus in Bellary district, Karnataka. Indian J Med Res 126:152–157

Kanojia PC, Shetty PS, Geevarghese G (2003) A long-term study on vector abundance & seasonal prevalence in relation to the occurrence of Japanese encephalitis in Gorakhpur district, Uttar Pradesh. Indian J Med Res 117:104–110

La casse, WJ, Yamaguti S (1955) Mosquito fauna of Japan and Korea. Off. Surg. 8th U.S. Army, Kyoto, Honshu

Mani TR, Mohan Rao CVR, Rajendran R, Devaputra M et al (1991) Surveillance for Japanese encephalitis in villages near Madurai, Tamil Nadu, India. Trans R Soc Trop Med Hyg 85:287–291

Menon PKB, Rajagopalan PK (1976) A note on *Culex tritaeniorhynchus* Giles, 1901, in villages near Delhi. Indian J Med Res 64:709–712

Muangman D, Edelman R, Sullivan MJ, Gould DJ (1972) Experimental transmission of Japanese encephalitis virus by *Culex fuscocephala*. Am J Trop Med Hyg 21:482–486

Pant U, Ilkal MA, Somen RS, Shetty PS et al (1994) First isolation of Japanese encephalitis virus from the mosquito *Culex tritaeniorhynchus* Giles, 1901 (Diptera: Culicidae) in Gorakhpur District, Uttar Pradesh. Indian J Med Res 99:149–151

Panwar RS, Kumar N (2008) *Cassia occidentalis* toxicity causes recurrent outbreak of brain disease in children in Saharanpur. Indian J Med Res 127:413–414

Peiris JSM, Amerasinghe FP, Amerasinghe PH, Ratnayaka CB et al (1992) Japanese encephalitis in Sri Lanka—a study of an epidemic: vector incrimination, porcine infection and human disease. Trans R Soc Trop Med Hyg 86:307–331

Philip Samuel P, Hiriyan J, Thenmozhi V, Balasubramanian A (1998) A system for studying vector competence for Japanese encephalitis virus. Indian J Malariol 35:146–150

Philip Samuel P, Hiriyan J, Gajanana A (2000) Japanese encephalitis virus infection in mosquitoes and its epidemiological implications. ICMR Bull 30:37–43

Reuben R (1971a) Studies on the mosquitoes of North Arcot District, Madras state, India. Part 1. Seasonal densities. J Med Entomol 8:119–126

Reuben R (1971b) Studies on the mosquitoes of North Arcot District, Madras state, India. Part 5. Breeding places of the Culex vishnui group of species. J Med Entomol 8(4):363–366

Reuben R (1971c) Studies on the mosquitoes of North Arcot District, Madras state, India. Part 6. Seasonal prevalence of the Culex vishnui group of species. J Med Entomol 8(4):367–371

Reuben R, Tewari SC, Hiriyan J, Akiyama J (1994) Illustrated key to genera of Culex (*Culex*) associated with Japanese encephalitis in Southeast Asia (Diptera: Culicidae). Mosq Syst 26:75–96

Reuben R, Thenmozhi V, Phillip Samuel P, Gajanana A et al (1992) Mosquito blood feeding patterns as a factor in the epidemiology of Japanese encephalitis in Southern India. Am J Trop Med Hyg 46:654–663

Rosen L, Tesh RB, Lien JC, Cross JH (1978) Trans-ovarian transmission of Japanese encephalitis virus by mosquitoes. Science 199:909–911

Sabesan S, Konuganti HKJ, Perumal V (2008) Spatial delimination, forecasting and control of Japanese encephalitis: India—a case study. Open Parasitol J 2:59–63

Saxena SK, Mishra N, Saxena R, Singh M et al (2009) Trend of Japanese encephalitis in North India: evidence from thirty-eight acute encephalitis cases and appraisal of niceties. J Infect Dev Ctries 3(7):517–530

Sirivanakarn S (1976) Medical entomology studies III. A revision of the subgenus *Culex* in the Oriental region (Diptera: Culicidae). Contrib Am Entomol Inst (Ann Arbor) 12(2):1–272

Thenmozhi V, Rajendran R, Philip Samuel P, Hiriyan J et al (2001) Natural vertical transmission of Japanese encephalitis in south Indian mosquitoes. Trop Biomed 18:19–27

Index

B. P. Das, *Mosquito Vectors of Japanese Encephalitis Virus from Northern India*, SpringerBriefs in Animal Sciences, DOI: 10.1007/978-81-322-0861-7, © The Author(s) 2013